William H. Birkmire

Compound riveted Girders

As applied in the Construction of Buildings

William H. Birkmire

Compound riveted Girders
As applied in the Construction of Buildings

ISBN/EAN: 9783337106379

Printed in Europe, USA, Canada, Australia, Japan

Cover: Foto ©ninafisch / pixelio.de

More available books at **www.hansebooks.com**

COMPOUND RIVETED GIRDERS,

AS APPLIED IN THE

CONSTRUCTION OF BUILDINGS.

WITH NUMEROUS

PRACTICAL ILLUSTRATIONS AND TABLES.

BY

WILLIAM H. BIRKMIRE,

AUTHOR OF "ARCHITECTURAL IRON AND STEEL" AND
"SKELETON CONSTRUCTION IN BUILDINGS."

NEW YORK:
JOHN WILEY & SONS,
53 EAST TENTH STREET.
1893.

COPYRIGHT, 1893,
BY
WILLIAM H. BIRKMIRE.

PREFACE.

IN order to facilitate the calculation attending the construction of Wrought Iron and Steel Riveted Girders, the author has endeavored in this work to supply the link which separates Theory from Practice. Its object may be briefly stated. A riveted girder is to be designed; the span, depth, and loads are known, the strains are calculated by the well-known bending-moment formulæ, and largely by the graphic method; lastly, the details of construction are fully illustrated.

Touching the question of accuracy, it is scarcely necessary to notice the slight difference that may arise between the two methods, i.e., working out the usual formulæ, or by measuring from the graphic diagrams. The time consumed in wading through a complicated series of equations to reach a few measurements is objectionable when at least such measurements can at once be had by the graphic method.

This work does not investigate exceptional or extremely scientific riveted girders, but more especially those of a type now extensively adopted and constructed by well-known architectural iron workers.

The diagrams and the various examples explaining the Author's method are submitted to architects and architectural students with the hope that they will become a medium of usefulness to them in the routine of office work.

WILLIAM H. BIRKMIRE.

NEW YORK, 1893.

TABLE OF CONTENTS.

PART I.
THE STRAINS IN COMPOUND RIVETED GIRDERS.

	PAGE
Compound riveted girders described	1
Bending moments	2
Flanges	4
Shearing forces on the webs	5
Buckling of webs	7
Stiffeners	8
Riveting	8
Frict on of plates	10
Proportioning rivets	11
Rivets connecting webs with flanges	12
Spacing rivets according to strain produced in the flanges by the bending moments	15
Proportioning girders	16
Shearing and bearing resistance of rivets (Table)	16
Details of construction	17
Extract from the New York Building Law in relation to riveted girders	18
To calculate the approximate weight of girder before its dimensions are fixed	19
Splicing	20

PART II.
QUALITY OF MATERIAL.

Wrought-iron	21
Limit of elasticity of wrought-iron	21
Ultimate strength of wrought-iron	21
Rivet iron	21
Mild steel	21

	PAGE
Ultimate strength and elongation	22
Rivet steel	22
Painting	22

PART III.

EXAMPLE I.

Girder supporting a concentrated load at centre of span	23
Construction of flanges in a girder supporting a concentrated load at centre	25
Flanges reduced in area towards the supports in a girder supporting a concentrated load at centre	26
Webs proportioned in a girder supporting a concentrated load at centre of span	28
Stiffeners in a girder supporting a concentrated load at centre of span	28
Rivet spacing in a girder supporting a concentrated load at centre of span	29
Graphical representation of bending moments and shearing forces in a girder with a concentrated load at centre of span	30
List of material and details of a girder supporting a concentrated load at centre	32
Areas of angles with even legs (Table)	33
" " " " uneven legs (Table)	33
Sectional area in inches of rivet-holes in plates of various thicknesses (Table)	34
Gross area of plates of various thicknesses (Table)	35
Safe buckling value of web plates in wrought-iron (Table)	35
Shearing value of wrought-iron web plates (Table)	36
" " " steel web plates (Table)	37

EXAMPLE II.

Girder supporting one concentrated load not at centre of span	38
Construction of flanges in a girder supporting one concentrated load not at centre	40
Flanges reduced in area in a girder supporting one concentrated load not at centre	41
Webs proportioned in a girder supporting a concentrated load not at centre	42
Stiffeners in a girder with one concentrated load not at centre	43
Spacing of rivets in a girder with one concentrated load not at centre	43
Graphical representation of bending moments and shearing forces in a girder with one concentrated load not at centre of span	44
List of material and details of a girder supporting one concentrated load not at centre of span	46

EXAMPLE III.

	PAGE
Girder supporting a uniformly distributed load	47
Construction of flanges in a girder supporting a uniformly distributed load.	49
Flanges reduced in area towards the supports of a girder supporting a uniformly distributed load	49
Webs proportioned in a girder supporting a uniformly distributed load	50
Stiffeners in a girder supporting a uniformly distributed load	51
Spacing of rivets in a girder supporting a uniformly distributed load	51
Method of drawing parabolas	52
Parabola by the construction of a diagram	53
Graphical representation of bending moments and shearing forces in a girder supporting a uniformly distributed load	54
List of material and details of a girder supporting a uniformly distributed load	55

EXAMPLE IV.

Girder supporting two concentrated loads	56
Construction of flanges in a girder supporting two concentrated loads	58
Flanges reduced in area towards the supports in a girder supporting two concentrated loads	59
Webs proportioned in a girder supporting two concentrated loads	60
Stiffeners in a girder supporting two concentrated loads	61
Spacing of rivets in a girder supporting two concentrated loads	61
Graphical representation of bending moments and shearing forces in a girder supporting two concentrated loads	62
List of material and details of a girder supporting two concentrated loads.	64

EXAMPLE V.

Girder supporting two concentrated loads and a uniformly distributed load	65
Construction of flanges in a girder supporting two concentrated loads and a uniformly distributed load	67
Webs proportioned in a girder supporting two concentrated loads and a uniformly distributed load	68
Stiffeners in a girder supporting two concentrated loads and a uniformly distributed load	69
Spacing of rivets in a girder supporting two concentrated loads and a uniformly distributed load	69
Graphical representation of bending moments and shearing forces in a girder supporting two concentrated loads and a uniformly distributed load	70
Flanges reduced in area towards the supports in a girder supporting two concentrated loads and a uniformly distributed load by the funicular polygon	71
List of material and details of a girder supporting two concentrated loads and a uniformly distributed load	73

TABLE OF CONTENTS.

EXAMPLE VI.

	PAGE
Girder supporting three concentrated loads	74
Construction of flanges in a girder supporting three concentrated loads	76
Webs proportioned " " " " " "	77
Stiffeners " " " " " "	78
Spacing of rivets " " " " " "	78
Flanges reduced in area towards the supports in a girder supporting three concentrated loads	79
Graphical description of bending moments and shearing forces in a girder supporting three concentrated loads	80
List of material and details of a girder supporting three concentrated loads	82

EXAMPLE VII.

Girder supporting four concentrated loads	83
Construction of flanges in a girder supporting four concentrated loads	85
Webs proportioned " " " " " "	85
Stiffeners " " " " " "	86
Spacing of rivets " " " " " "	86
Flanges reduced in area towards the supports in a girder supporting four concentrated loads	87
Graphical representation of the bending moments and shearing forces in a girder supporting four concentrated loads	88
List of material and details of a girder supporting four concentrated loads	90

EXAMPLE VIII.

Steel girder supporting five concentrated loads	91
Determination of bending moments in a girder supporting five concentrated loads	93
Construction of flanges in a girder supporting five concentrated loads	94
Stiffeners " " " " " "	95
Spacing of rivets " " " " " "	95
Flanges reduced in area towards the supports in a girder supporting five concentrated loads	96
Girder fixed at one end and loaded with a concentrated load at the other, as a cantilever	97
Girder fixed at one end supporting a uniformly distributed load, as a cantilever	98
Girder fixed at one end supporting more than one load, as a cantilever	99
The relative strength of simple and cantilever girders; maximum vertical shear, bending moments, and deflection (Table)	99
Modulus of elasticity of wrought-iron and steel in riveted girder as compared with solid sections, as I-beams	99
Moment of Inertia for Rectangular sections	100

PART IV.

TABLES.

	PAGE
Average weight in pounds of a cubic foot of various substances	101
Weight of 100 rivets in pounds	104
Decimal equivalents for fractions of a foot	105
Number of U. S. gallons contained in circular tanks	106
Decimal equivalents for fractions of an inch	106
Weight per lineal foot of cast-iron columns	107
Weight of square cast-iron columns per lineal foot	108
Weight per foot of flat iron	109
Table of squares and cubes	111
Table of circles	115

Shearing and bearing resistance of rivets	16
Areas of angles with even legs	33
" " " " uneven legs	33
Sectional area in inches of rivet-holes in plates of various thicknesses	34
Gross area of plates of various thicknesses	35
Safe buckling value of web plates (wrought-iron)	35
Shearing value of wrought-iron web plates	36
" " " steel web plates	37

LIST OF ILLUSTRATIONS.

PART I.

FIG.		PAGE
1.	Plate-girder section..	1
2.	Plate-girder section with single flange plate.........................	1
3.	Plate-girder section with three flange plates.........................	1
4.	Box-girder section with three flange plates...........................	1
5.	Box girder-section with three webs.....................................	1
6.	Girder with two loads supported upon a fulcrum....................	3
7.	A simple girder supported at each end and load in middle.......	3
8.	A lever held up with a weight at either end..........................	5
9.	A simple girder with load out of centre................................	6
10.	A simple girder with a specified load out of centre................	6
11.	Two plates riveted with rivets in single shear......................	12
12.	Plate-girder section with rivets in double shear....................	12
12a.	Box-girder section with rivets in single shear......................	12
13.	Girder illustrating the strains on rivets connecting flange with web...	12

PART III.

14.	Diagram of a girder with one concentrated load at centre...........	23
15.	Diagram determining position of flange plates in a girder of one concentrated load at centre...	26
16.	Section of plate girder with stiffeners bent around chord angles......	29
17.	Section of plate girder with straight stiffeners and fillers............	29
18.	Diagram of the graphical representation of bending moments and shearing forces of a plate girder with one concentrated load at centre...	31
19.	Detail of girder of one concentrated load at centre.................	32
20.	Diagram of a girder with one concentrated load not at centre.......	38
21.	Diagram determining position of flange plates in a girder of one concentrated load not at centre..	41
22.	Diagram of the graphical representation of bending moments and shearing forces in a girder with one concentrated load at centre..	45
23.	Detail of girder of one concentrated load not at centre.............	46
24.	Diagram of a girder with a uniformly distributed load..............	47
25.	Diagram determining position of flange plates in a girder supporting a uniformly distributed load...	50

LIST OF ILLUSTRATIONS.

FIG.		PAGE
26.	Diagram of a parabola, by ordinates from a tangent to a parabola at its vertex.............................	
27.	Diagram of a parabola, by lines to two sides of an isosceles triangle..	53
28.	Diagram of the graphical representation of bending moments and shearing forces in a girder supporting a uniformly distributed load...........	54
29.	Detail of girder supporting a uniformly distributed load..............	55
30.	Diagram of a girder supporting two concentrated loads..............	56
31.	Diagram determining position of flange plate in a girder supporting two concentrated loads......................................	59
32.	Diagram of the graphical representation of bending moments and shearing forces in a girder supporting two concentrated loads.....	62
33.	Detail of girder supporting two concentrated loads...................	64
34.	Diagram of a girder supporting two concentrated loads and one uniformly distributed load..	66
35.	Diagram of the graphical representation of bending moments and shearing forces in a girder supporting two concentrated loads and one uniformly distributed load.................................	70
36.	Diagram determining position of flanged plates in a girder supporting two concentrated loads and a uniformly distributed load.........	72
37.	Detail girder supporting two concentrated loads and a uniformly distributed load...	73
38.	Diagram of a girder supporting three concentrated loads............	75
39.	Diagram determining position of flange plates in a girder supporting three concentrated loads	79
40.	Diagram of the graphical representation of bending moments and shearing forces in a girder supporting three concentrated loads...	80
41.	Detail of girder supporting three concentrated loads	82
42.	Diagram of a girder supporting four concentrated loads.............	84
43.	Diagram determining position of flange plates in a girder of four concentrated loads...	88
44.	Diagram of the graphical representation of bending moments and shearing forces in a girder supporting four concentrated loads....	89
45.	Detail of girder supporting four concentrated loads.................	90
46.	Diagram of a girder supporting five concentrated loads..............	92
47.	Diagram determining position of flange plates in a girder supporting five concentrated loads..	96
48.	Diagram of a girder secured at one end (as a cantilever) and supporting a concentrated load at the other.............................	98
49.	Section of a plate girder, determining the notation used in the calculation of the section for the moment of inertia....................	100
50.	Section of a box girder, determining the notation used in the calculation of the section for the moment of inertia....................	100

PART I.

THE STRAIN IN COMPOUND RIVETED GIRDERS.

PART I.

THE STRAINS IN COMPOUND RIVETED GIRDERS.

For buildings, as well as railway and highway bridges, there is probably no other form of girders more extensively used than those made up of plates and angles, called Compound Riveted Girders.

Some of the principal reasons for this lies mainly in the simplicity of their construction; they can be adopted for any load or number of loads, and accommodated to any span usually met with in the construction.

The single web or plate girder is more economical, more accessible for painting and inspection. Formed of a single web and four angles, as Fig. 1, suitable for light loads and short spans; for heavier loads a single plate is added to the top and bottom flanges, as shown in Fig. 2; for still heavier loads additional plates, as in Fig. 3.

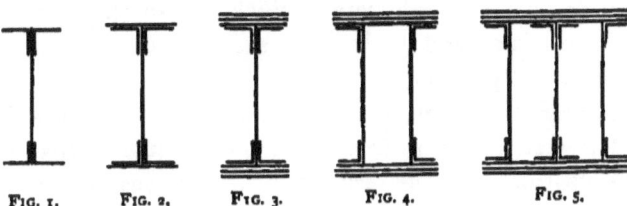

FIG. 1. FIG. 2. FIG. 3. FIG. 4. FIG. 5.

Where thick walls are to be supported and lateral stiffness is required, the double web or box girder, Fig. 4, or the triple

1

web, Fig. 5, is employed. It also becomes necessary in many cases to place two plate or two box girders side by side.

The box girders as represented in section, Fig. 4, is considered superior to the plate girders represented in Figs. 1, 2, and 3; but the preference should be given the latter on account of its simplicity of construction, and although inferior in strength to the box girder it has nevertheless other valuable properties to recommend it.

On comparing the strengths of these separate girders, weight for weight, it will be found that the box girder is as 1 to .93, or nearly as 100 to 90. The difference in strength does not arise from want of proportion in the top and bottom section of either girder, but from the position of the material; which in that of the box girder offers greatly superior powers of resistance to lateral flexure. The box girder, it will be observed, contains larger exterior sectional area, and is consequently stiffer and better calculated to resist lateral stress, in which direction the plate girder generally yields before its other resisting powers of *tension* and *compression* can be brought fully into action. Taking this girder, however, in a position similar to that in which it is used in supporting floor-beams and floor-arches of buildings, its strength is very nearly equal to that of the box shape, and, as previously mentioned, is of more simple construction, less expensive, and more durable, from the circumstance that the web-plate is thicker than the web-plates of the box girder, and it admits of easy access to all its parts for purposes of painting, etc.

Bending Moments.—Generally, the strength of a compound riveted girder is founded on the equality that must always exist between the resultant of the various loads tending to cause its rupture and the strength of the material of which the girder is composed. The former may be resolved horizontally into strains, depending for their value upon what are known as *moments of rupture*, *bending moments*, or *leverage*, of greater or less complexity, tending to cause the failure of the girder by tearing asunder its fibres in the bottom flange, crushing them together in the top flange, and vertically upon the web into what are known as shearing forces, due to the transmission of the vertical pressure of the loads to the points of support.

The strain produced in the flanges is resisted by a leverage equal to the depth of the girder, that is, between the centre of gravity of the flanges, and the amount per square inch of section with which the metal may be safely trusted.

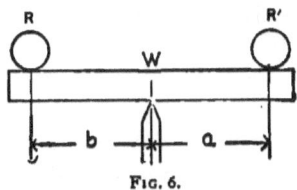

Fig. 6.

The bending moment is a compound quantity resulting from the multiplication of a force by a distance, and designated by the letter M. The forces are expressed in tons or pounds, and the distances in feet or inches; then the bending moments are in ton-feet or pound-inches.

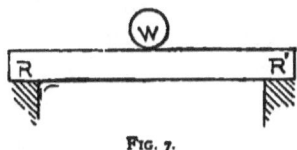

Fig. 7.

If b or a is the arm of leverage, Fig. 6, and a load R or R' acts for a distance from W, M at W is equal to the load R or R' multiplied by the distance b or a.

Then $M = Rb$ or $R'a$.

The direction of the stresses upon the girder are vertical, those at the ends being downwards, while that at the middle is upwards. In Fig. 7 we have a girder supported at both ends, and a load W resting upon the middle of its length. Comparing this with Fig. 6, we see that the stresses here are also vertical,* but in reversed order, the one at the middle being downwards, while those at the end are upwards. In other respects we have the same conditions as in Fig. 6.

M does not represent strain, being independent of depth, but is converted into flange strain by dividing by the depth; the strain then found, divided by the maximum unit strain, determines the number of square inches to be given to the flanges.

The maximum unit strain herein adopted is 6 tons (12,000 pounds) per square inch for wrought-iron and 7 tons (14,000 pounds) for steel.

Here $A =$ Area of flange $d =$ depth in feet, $s =$ unit strain in tons, and $M =$ bending moment. Then

$$A = \frac{M}{ds}.$$

Flanges. —Compound girders are unlike rolled beams, in which every fibre is connected; but have strains transmitted only through rivets which are distributed only at certain distances apart; consequently the flange angles are at every point more or less subjected to strains in addition to their own. This additional strain will evidently increase with the amount of plates. It is good practice, therefore, to make the girder so deep that the flanges do not require a number of plates to be packed one upon another, and then to choose angles as heavy as possible consistent with the total flange area required.

---

* "We have to distinguish between the outer forces which may act at various portions of the girder tending to cause motion of its parts, and the inner forces which prevent this motion. The first we may call stresses, and the second strains. We therefore speak of the 'stresses' *upon* a girder and the 'strains' *in* a girder."

In order to give the single-web girders the greatest amount of resistance, it is usual to use angles with unequal legs with the longer leg horizontal.

It sometimes becomes important to have the plates of the top flange extend from end to end, even when angles may be found which alone are sufficient to make up the required section, as it gives great lateral stiffness to the flange, and also helps to distribute the stress more uniformly than with the angles alone.

In box girders the flange plate adjoining the angles are required to extend from end to end.

In making up the bottom flange, rivet-holes must be deducted to obtain the net section, and in so doing the diameter of the rivet-hole should be taken at least ⅛ inch larger; this latter provides to a certain extent for the damage done to the strength of the metal in the process of punching or drilling.

For the top flange the gross sectional area may be taken as making up the same, providing the riveting is well done, i.e., the rivet completely filling up its own hole.

Shearing Forces on the Webs.—It is by the law of the lever that we are enabled to determine precisely what portion of a given load resting upon a girder is sustained by either point of support; the loads balancing each other at either end

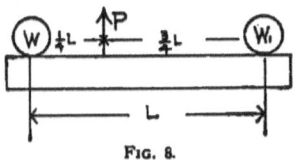

Fig. 8.

of a girder, or lever, on any point are to each other inversely as their distances (called lever arms) from the point or fulcrum.

For example: suppose we have a girder held up as in Fig. 8, with a load at either end, the point of support being to

COMPOUND RIVETED GIRDERS.

one side of the centre, say one-fourth of the lever arm from one end. In order that the lever be balanced, the load at W' must be one-fourth the sum of W and W', and that at W three-fourths that sum, for W' multiplied by $\frac{3}{4}L$ must always equal W multiplied by $\frac{1}{4}L$, and the sustaining force P must of course equal the sum of W and W'.

Again: supposing that there is one load as in Fig. 9. This

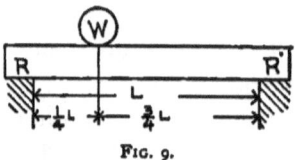

Fig. 9.

condition is the same as before, only reversed; and, according to the law of the lever, we find that for equilibrium a force must be applied to R equal to $\frac{3}{4}W$. This example is precisely the same as that of a girder, only R and R' are now called *reactions* of the supports, the sum of which must always be equal to the load or number of loads causing them.

In order, then, to know just how much of the load or number of loads at any point of the girder is supported by either support, all that is necessary to be done is to multiply the shorter or longer distance by the load and divide the product by the span L.

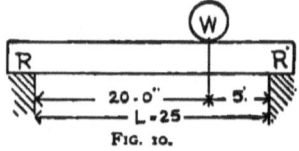

Fig. 10.

Example: Suppose we have a girder R and R', Fig. 10, of 25 feet span, and there is a load of 20 tons 5 feet from R'. Then each sustains a certain amount of this load proportionately to its distance from the load, the sum of the reactions being equal to the load,

$$R \text{ supports } \frac{20 \times 5}{25} = 4 \text{ tons};$$

$$R' \quad " \quad \frac{20 \times 20}{25} = 16 \text{ tons}.$$

The most practical way of proportioning the web is then to make its section sufficient to resist this entire shearing force at either end of the girder. Under the supposition that the flanges alone resist the entire bending moment, and the web only the shearing action, the following formula can be adopted:

Let S = shearing stress;

A = area at point of stress;

K = effective resistance to shearing;

t = thickness of web;

d = depth of web in inches.

$$A = td \quad \text{and} \quad S = ktd, \quad \text{or} \quad t = \frac{S}{dK}.$$

The safe shear on the webs per square inch herein adopted is 6000 lbs. for wrought-iron and 7000 lbs. for steel.

Example.—Suppose we take the above wrought-iron plate girder, Fig. 10, and have 16 tons (32,000 lbs.) shear on the web at R' support, the web being 12 inches in depth.

$$t = \frac{S}{dk} = \frac{32000}{12 \times 6000} = .44\text{—more than } \tfrac{7}{16} \text{ of an inch in}$$

thickness.

Buckling of Web.—The web is still in danger of *buckling* under this compression stress; consequently the web with its thickness as already proportioned for shearing must now be examined for its strength as a column. The depth being the vertical distance between the upper and lower rows of rivets in the web (but to facilitate the calculation the height will be taken the full depth of web). This condition is attained when the shear per square inch of cross-section at any point does not exceed

$$\text{The safe resistance to buckling per square inch} = \frac{10000}{1 + \frac{d^2}{3000 t^2}},$$

when d and t are the depth and thickness of web in inches.

Stiffeners.—If the result obtained by the above formula is less than that adopted by our safe shear, vertical angles or *stiffeners* are riveted each side of the web at intervals. They should always be used at the bearings and where concentrated loads occur.

The spacing of stiffeners is more a matter of experimental judgment than of mere calculation. Of several rules given by engineers, one is, "that stiffeners in girders over three feet in depth shall be placed at distances apart (centre to centre) generally not exceeding the depth of the full web-plate, with the maximum limit of 5 feet."

In girders under 3 feet in depth stiffeners may be placed 3 feet apart, and in some special cases where there is little or no shearing, at greater distances. This refers, of course, to those parts of the girders where there are no concentrated loads.

Another rule worthy of notice says "that when the least thickness of web is less than $\frac{1}{80}$ the depth of the girder, the web shall be stiffened at intervals not over twice the depth."

Riveting.—The rivets in girder work are generally the same size as those adopted for boiler work, i.e., $\frac{3}{4}''$, $\frac{7}{8}''$, and $1''$ in diameter; but as girders do not require caulking like a boiler, the pitch or distance of rivets from centre to centre is much greater, and usually varies from $2\frac{1}{2}$ diameters to 16 times the thickness of the outside plate joined. By diameters is understood the diameter of the shank. When the edges of the plate are often roughly shorn, the margin, or distance between the rivet-hole and edge of the plate, is seldom less than $1\frac{1}{4}$ times the diameter of the rivet. As the effect of punching is to weaken the plate some distance all round the punched hole, the above proportions should be adopted.

Nearly all experimenters on the subject agree that punching generally reduces the tenacity of iron and steel plates to a greater degree than the area of the metal punched out, and a close examination of the border of each hole shows that it has been subject to a certain degree of rupture, which in most cases has reduced the ductility of the metal and made it wholly crystalline in fracture, and as some may suppose caused cracks round the edge of the hole; but this latter seems doubtful, as Mr. Gerhard (see Experiments in Stoney's Riveted Joints) instituted an investigation as to whether there was any foundation for the very generally received opinion that the edges of a punched hole on the *die* side are injured by a ring of minute incipient *cracks*. For this purpose a large number of specimens 5 inches by 3 inches by $\frac{1}{2}$ inch of all kinds of steel were prepared. The edges were planed, the surfaces polished, holes were pierced in various ways, and the metal surrounding them was carefully examined with a microscope, but no trace whatever of *cracks* was found, though the nature of the steel ranged from 0.1 to 0.6 per cent of carbon. Owing to its hardness and inability to stretch, this annulus of strained material round the punched holes, when the specimen is under tension, takes a higher proportion of the stress than the other more yielding parts, and hence it reaches the breaking point sooner, that is, the punched plate breaks in detail: first the annulus gives way, and then the more ductile portion between the holes. Reaming or boring out a zone of metal $\frac{1}{8}$ inch wide round the punched hole removes the annulus of strained material and neutralizes the effect of punching. In numerous experiments on the subject the loss of tenacity in iron plates from punching varies from 5 to 23 per cent of the original strength of the solid plate, but the percentage in any particular case will doubtless depend (1st) on the diameter of the holes; (2d) on the pitch; (3d) on the width of the strip punched, for wide plates are apparently less injured than narrow strips; (4th) on the condition of the punching tool—i.e., the sharpness of its cutting edges—and the maintenance of the proportion of size between the *punch* and the *die;* (5th) on the quality and thickness of the metal, hard iron generally suffering more than ductile iron, and thin plates less than thick ones. Probably the most accurate method for making an allowance for the injurious effect of punching is to allow a certain percentage when calculating the effective net area of a punched plate, and, as heretofore mentioned under Flanges, page 5, $\frac{1}{8}$ of an inch more than the diameter of the rivet is adopted.

Friction of Plates.—Rivets contract in cooling and draw the plates together with such force that the friction produced between their surfaces is generally sufficient to prevent them from sliding over each other so long as the stress lies within limits which are not exceeded in ordinary practice.

The friction of plates is an important factor in boiler work; and as it is usual, to test them hydraulically, to double their working pressure, the joints are so designed that this water test, as well as the expansion and contraction due to changes in temperature, will not cause the joints to slip. Though the friction of riveted plates may be sufficient to convey the normal working load without subjecting the rivets to a shearing stress, it does not follow, nor do experiments indicate, that the *ultimate* strength of a riveted joint is increased by this friction.

When several plates are riveted together with numerous rivets, as in the piled flanges of a girder, the slipping of plates does not seem to have occurred in Mr. Baker's experiments, for with two wrought-iron girders with 5 and 8 plates, respectively, in their flanges, each 20 feet span and 2 feet in depth, which he tested to failure, there was no movement in the flanges, and the pile of plates behaved almost like a welded mass of iron, and Mr. Baker states "that he invariably found that badly punched girders, with the holes partly blind and the rivets tight but not filling the holes, deflected neither more nor less than the most accurately drilled work."

Whatever value may justly be attached to the above, an inspection of the riveted joint when being tested to destruction dispels all idea of the ultimate stress being in any degree affected by it; for when the stress is considerable, the joints open at each end of the plates, and the higher the stress the greater the amount of opening is observed. Under such conditions it is not customary to take into consideration the friction of the joints.

Proportioning Rivets.—Rivets, as used for girders, must be proportioned to resist *shearing*, and the area of their bearing must be such that the metal against which they bear shall not be crushed. The stresses allowed on these members are: shearing, 7500 to 9000 pounds, and crushing, 15,000 pounds per square inch.

The shearing strain is measured on the area of the cross-section of the rivet; the crushing, on the area obtained by the product of the diameter of the rivet by the thickness of the *web* or plate upon which it bears.

To illustrate the shearing and bearing area of a rivet, we take for example two plates of wrought-iron 8 inches wide by $\frac{1}{2}$ inch thick, which overlap each other for a joint, with 45,000 pounds strain on the plates; what number of rivets will be required to resist the strain on the joint?

The area of a rivet $\frac{3}{4}$ of an inch in diameter is 0.4417 square inches; this multiplied by 7500 pounds, the safe shearing, = 3312.75 pounds, the safe amount of strain each rivet can sustain without shearing; dividing 45,000 by this, we get 13.6, say 14 rivets, as shown in *single shear*, Fig. 11. If constructed as shown in Fig. 12, as in the flange of a plate girder, the rivets would be in *double shear* and have twice the value; then 7 rivets would be sufficient.

Fig. 11.   Fig. 12.   Fig. 12A.

In box girders, as Fig. 12*a*, the rivets connecting the angles with the webs are in single shear.

The bearing area of each rivet is ¾ inch by ½ inch = ⅜ square inch; this multiplied by 15,000 pounds for crushing would equal 5625 pounds. Dividing 45,000 by this we obtain 8 rivets. This latter calculation should not be overlooked in riveted work. Its observance in most cases of riveted girders with single webs gives the size and number of rivets to be used, and in thin webs the bearing area may be small, necessitating a thicker web than would otherwise be required.

Rivets Connecting Web with Flanges. — The strain which the rivets connecting the web and flanges sustain is evidently due to the strain which is transmitted from one to the other; this strain is horizontal, and is the maximum increment of flange strain at every section of the girder, and is found by dividing the *maximum shear* at any point by the height of the girder.

For example: suppose a girder of 24 feet span (Fig. 13), 3 feet in depth, sustains a load of 150,000 pounds uniformly distributed over its whole length. Taking half the load over half the girder, at the support R' the shearing strain is 75,000 pounds or half the whole load; at *a* or 3 feet it is equal to ¾ of half the load or 56,250 pounds; at *b* or 6 feet it is equal to ½ of half the load or 37,500 pounds; at *c* or 9 feet it is equal to ¼ of half

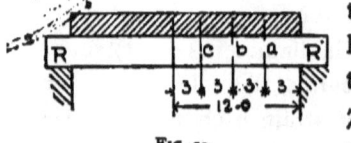

Fig. 13.

THE STRAINS IN COMPOUND RIVETED GIRDERS. 13

the load or 18,750 pounds; at 12 feet or the centre it is zero: from which we can now obtain the flange strain by dividing by the height, and again by 12 inches to get the pounds per inch of run.

Then at end $= \dfrac{75000}{3 \times 12} = 2083$ lbs. per inch of run;

at $a = \dfrac{56250}{3 \times 12} = 1562$ " " " " "

at $b = \dfrac{37500}{3 \times 12} = 1041$ " " " " "

at $c = \dfrac{18750}{3 \times 12} = 520$ " " " " "

The result thus obtained is then for the shear on the rivets. If the girder has a single web, as in a plate girder, we will take its bearing value, using a $\frac{3}{8}$ web and a $\frac{7}{8}$-diameter rivet.

The value of each rivet would be $15{,}000 \times \frac{7}{8} \times \frac{3}{8} = 4920$ pounds. Then at the end where we have the stress of 2083 pounds per inch run, we must therefore space rivets:

at support $= \dfrac{4920}{2083} = 2\frac{3}{8}$ inches centre to centre;

at $a = \dfrac{4920}{1562} = 3\frac{1}{8}$ " " " "

at $b = \dfrac{4920}{1041} = 4\frac{3}{4}$ " " " "

at $c = \dfrac{4920}{520} = 9\frac{7}{16}$ " " " "

But as we have exceeded our limit at c, we will require the spacing from c to middle of girder to be 6 inches centres.

COMPOUND RIVETED GIRDERS.

In order that rivets in the two legs of the angle should stagger each other, both legs must have the same spacing; and in order that the rivets may lie in straight lines vertical and horizontal, the top and bottom rows should be spaced alike. In fact, the spacing in top and bottom flanges for practical considerations are similar.

Generally the vertical in addition to the horizontal strain is taken into consideration for spacing the rivets, and their resultant is therefore the strain on the rivets. The vertical strain is that due directly to the load resting upon the flange of the girder, and thence through the rivets transmitted to the web.

In the above example, Fig. 13, the flange strain at the end

$$= \frac{75000}{3 \times 12} = 2083 \text{ lbs. per inch of run.}$$

The vertical load on the girder per inch of run is the total load, divided by the length of span for the load on one foot, and the quotient by 12 inches for the per inch of run of load,

$$\text{or } \frac{150000}{24 \times 12} = 520 \text{ lbs. per inch of run.}$$

Then from the above at end of girder the *resultant*

$$= \sqrt{2083^2 + 520^2} = 2147 \text{ lbs. per inch of run.}$$

The spacing of rivets at end of girder $= \frac{4920}{2147} = 2\frac{5}{16}$ inches.

At a, 3 feet from end, flange strain $= 1562$ lbs. per inch of run.

The vertical stress as given is 520 lbs. per inch of run.

The resultant $= \sqrt{1563^2 + 520^2} = 1647$ lbs. per inch of run.

Then $\frac{4920}{1647} = 3$ inches, the spacing of rivet at 3 feet from end of girder. Continue in like manner to the centre.

Spacing Rivets according to Strain Produced by the Bending Moments.—The rivets are also spaced, in the angles which connect the flanges with the web, according to the strain produced in the flanges by the bending M, and the number readily found with but little calculation. The horizontal strain in the flanges diminishes in intensity either way from the position of maximum M towards either support, where it is the least, and may be found (as mentioned before under Bending Moments) by dividing by the depth.

The horizontal increments of strain in the web are greater, however, at the ends, and least under position of maximum M. If the maximum strain in the flange is divided by the value of each rivet, there results the minimum number of rivets either way from maximum M to either support.

For example: in a girder of 24 feet span and 3 feet in depth, $\frac{3}{8}$ web and $\frac{7}{8}$ rivets, if the maximum M equals 225 ton-feet at centre of girder, and the flange stress is at that point $2\frac{2}{3}\frac{5}{3} = 75$ tons, or 150,000 lbs., the number of rivets

$$= \frac{150,000}{4920} = 30 \text{ required either way from centre}$$

a distance of 144 inches, or spaced $\frac{144}{30} = 4\frac{12}{30}$ inches. Owing, however, to the greater intensity of the horizontal increment of strain in the web towards these supports, the rivets should be spaced closer as the ends are approached.

Then for the first 3 feet, say 3 inches centres; the next, 4 inches; the next, 5 inches; and the remaining 3 feet, 6 inches. Then there results a total of 34 rivets.

For any further explanation or spacing of rivets refer to the various examples which follow.

For convenience in selection of rivets, the following table has been prepared:

SHEARING AND BEARING RESISTANCE OF RIVETS.

Diameter of rivet in inches.		Area of rivet in square inches.	Single shear at 7500 lbs. per square inch.	Bearing resistance in pounds for different thickness of plates at 15,000 lbs. per square inch.								
Fraction.	Decimal.			¼	5/16	⅜	7/16	½	9/16	⅝	11/16	¾
⅝	0.625	0.3068	2300	2340	2930	3520						
¾	0.75	0.4418	3310	2810	3510	4220	4920	5630	6330			
⅞	0.875	0.6013	4510	3280	4100	4920	5740	6562	7380	8200	9020	9840
1	1	0.7854	5890	3750	4690	5620	6562	7500	8440	9380	10310	11250

Proportioning Girders.—The first operation, that of obtaining the kind of girder, is not always left entirely to the discretion of the designer; no rules can be laid down, for the reason that various loads on fixed spans and depths are given, so that, in the nature of the construction, very little limit is allowed; for instance, if the girders are in the floor construction, the height of girder is reduced to a minimum, to give the greatest height of ceiling.

The depth at centre of straight independent girders as given by Humber may be made from $\frac{1}{10}$ to $\frac{1}{16}$ of the span. The greatest economy of material is perhaps obtained at $\frac{1}{12}$.

If the depth of girder is about $\frac{1}{16}$ the span, the *deflection* will not be too great.

For many cases it would be well to find the most economical depth by a few trials, and bearing in mind that the increase of depth decreases the flange area, while it increases the weight of web and stiffeners and *vice versa*.

It has been previously mentioned that the depth of the girder is between the *centre of gravity of the flanges*, where we have one or more plates. We can without much error assume the distance between the centre of gravity of the flanges to be equal to the *distance from top to bottom of flange-angles* as the *effective depth* of the girder. As the flange section increases the effective depth increases, but we can assume them to be constant throughout.

THE STRAINS IN COMPOUND RIVETED GIRDERS. 17

The span should include the length between the centre of bearings or supports, but for all practical purposes the *effective* span herein taken is between the supports.

The following general rules should be adopted in proportioning girders.

1. Plate girders should be proportioned upon the supposition that the bending or chord strains are resisted entirely by the upper and lower flanges, and that the shearing or web strains are resisted entirely by the web plate.

2. In members subject to tensile strains, full allowance shall be made for reduction of section by rivet-holes, etc.

3. The web plates shall not have a shearing strain greater than 6000 to 8000 pounds for wrought iron and 7000 to 9000 pounds for steel per square inch, and no web plate shall be less than $\frac{3}{8}$ inch in thickness.

4. No wrought-iron or steel shall be used less than $\frac{3}{8}$ inch thick, except in places where both sides are always accessible for cleaning and painting.

DETAILS OF CONSTRUCTION.

1. All the connections and details of the several parts shall be of such strength that, upon testing, rupture shall occur in the body of the members rather than in any of their details or connections.

2. The webs of plate girders, when they cannot be had in one length, must be spliced at all joints by a plate on each side of the web. T-iron must not be used for splices.

3. When the least thickness of the web is less than $\frac{1}{80}$ of the depth, the web shall be stiffened at intervals not over twice the depth of the girder.

4. The pitch of rivets shall not exceed 6 inches, nor sixteen times the thinnest outside plate, nor be less than three diameters of the rivet in a straight line.

5. The rivets used will be generally $\frac{3}{4}$ and $\frac{7}{8}$ inch diameter.

6. The distance between the edge of any piece and the centre of a rivet-hole must never be less than $1\frac{1}{4}$ inches.

7. In punching plates or other iron, the diameter of the die shall in no case exceed the diameter of the punch more than $\frac{1}{16}$ of an inch.

8. All rivet-holes must be so accurately punched that, when the several parts forming one member are assembled together, a rivet $\frac{1}{16}$ inch less in diameter than the hole can be entered, hot, into any hole without reaming or straining the iron by "drifts."

9. The rivets when driven must completely fill the holes.

10. The rivet-heads must be hemispherical, and a uniform size for the same sized rivets throughout the work. They must be full and neatly made, and be concentric to the rivet-hole.

11. Whenever possible, all rivets must be machine-driven.

12. The several pieces forming one built member must fit closely together, and, when riveted, shall be free from twists, bends, or open joints.

13. All joints in riveted work, whether in tension or compression members, must be fully spliced, as no reliance will be placed upon abutting joints. The ends, however, must be dressed straight and true, so that there shall be no open joints.

14. All bed-plates under bearings of girders must be of such dimensions that the greatest pressure on the masonry shall not exceed 250 pounds per square inch.

EXTRACT FROM THE NEW YORK BUILDING LAW
PASSED APRIL 9, 1892.

§ 486. "Rolled iron or steel beam girders, or *riveted* iron or steel plate girders used as lintels or as girders, carrying a wall or floor or both, shall be so proportioned that the loads which may come upon them shall not produce strains in tension or compression upon the flanges of more than 12,000 pounds for iron nor more than 15,000 pounds for steel per square inch of the gross section of each of such flanges, nor a shearing strain upon the web plate of more than 6000 pounds per square inch of section of such web-plate if of iron, nor more than 7000 pounds if of steel; but no web plate shall be less than one quarter of an inch in thickness. Rivets in plate girders shall not be less than ⅝ of an inch in diameter, and shall not be spaced more than 6 inches apart in any case. They shall be so spaced that their shearing strains shall not exceed 9000 pounds per square inch of section, nor their bearing exceed 15,000 pounds per square inch, on their diameter, multiplied by the thickness of the plates through which they pass. The riveted plate girders shall be proportioned upon the supposition that the bending or chord strains are resisted entirely by the upper and lower flanges, and that the shearing strains are resisted entirely by the web plate. No part of the web shall be estimated as flange area, *nor more than one half of that portion of the angle-iron which lies against the web.*

The distance between the centre of gravity of the flange areas will be considered as the effective depth of the girder.

Before any girder, as before mentioned, to be used in any building shall be so used, the architect or the manufacturer or a contractor for it shall, if required so to do by the superintendent of buildings, submit for his examination and approval a diagram showing the loads to be carried by said girder, and the strains produced by such load, and also showing the dimensions of the materials of which said girder is to be constructed to provide for the said strains; and the manufacturer or contractor shall cause to be marked upon said girder, in a conspicuous place, the weight said girder will sustain, and no greater weight than that marked on such girder shall be placed thereon."

To Calculate the Approximate Weight of Girder before its Dimensions are Fixed.—It should be remarked here that the weight of the girder becomes considerable when the flanges are built up of a number of plates. It is therefore desirable to be able to calculate approximately the weight of the girder before its dimensions have been definitely fixed. The weight of the girder will be in proportion to its area of cross-section and to its length; or when W is the gross load to be carried, and L the length between the supports, then the weight of girder between the bearings is

$$w = \frac{WL}{C},$$

in which C is a constant, and W the load to be supported. The value of C has been taken from examples of girders from 35 to 50 feet long; its value is found to be 700.

Example: We have a span of 40 feet and 70 tons to be supported; what will be the approximate weight of girder?

$$w = \frac{WL}{C} = \frac{70 \times 40}{700} = 4 \text{ tons},$$

making a total of 74 tons, uniformly distributed.

Splicing.—Girders 40 feet and less will not require any splicing, as the plates and angles can be readily handled in one length.

In splicing the top flange, no additional cover plate will be required over the joint, but the ends should be planed true and butt solidly. The rivets to be closer near the joint.

The plate covering the joint of bottom flange requires to be the same area as the plates joined, and of sufficient length to take a number of rivets equal to strength of the cover plate.

PART II.

QUALITY OF MATERIAL.

PART II.

QUALITY OF MATERIAL.

Wrought-iron.—All wrought-iron must be tough, fibrous, and uniform in character. It shall have a limit of elasticity of not less than 26,000 pounds per square inch. The tensile strength, limit of elasticity, and ductility shall be determined from a standard test-piece about $\frac{1}{2}$ square inch. The elongation shall be measured on an original length of 8 inches.

When taken from plates rolled to a section of not more than $4\frac{1}{2}$ square inches, the iron shall show a minimum ultimate strength of 50,000 pounds per square inch, and a minimum elongation of 18 per cent in 8 inches. The same sized specimen, taken from plates 8 inches to 24 inches in width, shall show a minimum ultimate strength of 48,000 pounds per square inch, and a minimum elongation of 15 per cent in 8 inches; plates from 24 inches to 36 inches wide, 46,000 pounds per square inch, and elongate 10 per cent in 8 inches; plates over 36 inches wide, 8 per cent in 8 inches.

The same sized specimen taken from angle-iron shall have a minimum ultimate strength of 48,000 pounds per square inch, and a minimum elongation of 15 per cent in 8 inches. Rivet-iron shall have the same physical requirements as high-test iron, and in addition shall bend cold 180 degrees to a curve whose diameter is equal to the thickness of the rod tested, without signs of fracture on the convex side.

All iron for tension members must bend cold through 90 degrees to a curve whose diameter is not over twice the thickness of the piece, without cracking; at least one example in three must bend through 180 degrees to this curve without cracking. When nicked on one side and bent by a blow from a sledge, the fracture must be nearly fibrous.

Mild Steel.—Specimens from finished material for test, cut to size, as for wrought-iron, shall have an ultimate strength of from 54,000 to 62,000 pounds per square inch, with a minimum elongation of 26 per cent in 8 inches; to bend cold 180 degrees flat on itself, without sign of fracture on the outside of bent portion.

All rivets of mild steel must, under the above bending test, stand closing solidly together without sign of fracture.

Painting.—All iron and steel work, before leaving the shop, shall be thoroughly cleansed from all loose scale and rust, and be given one good coating of best oxide of iron and pure linseed oil, and after erection to receive one additional coat of paint.

PART III.

EXAMPLES.

PART III.

EXAMPLE I.

GIRDER SUPPORTING A CONCENTRATED LOAD AT CENTRE OF SPAN.

In a girder supported at both ends with a load concentrated at centre of span, the *maximum* bending moment is at the centre, and equals half the load multiplied by half the span, or

$$M = \frac{WL}{4}.$$

To find the bending moment at any point in the girder when the load is at the centre.

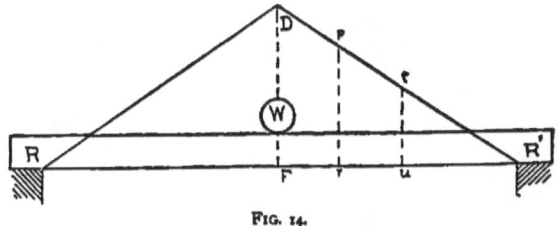

FIG. 14.

Make FD, by any scale, Fig. 14, equal M at centre of span; join RD and DR'. Then by the same scale, rp will equal the

COMPOUND RIVETED GIRDERS.

bending moment at point r, and ut will equal the moment at point u. Or the moment at any point r or u will equal half the load multiplied by $R'r$ or $R'u$; R' being the nearest support. Then

$$\text{At } r, \quad M = \frac{W}{2} R'r.$$

$$\text{At } u, \quad M = \frac{W}{2} R'u.$$

Example: What metal area would be required in the flanges at the centre of a girder of 30 feet span, 2 feet in depth, to sustain 40 tons concentrated at the centre of span; 6 tons (12,000 pounds) being the maximum unit strain per square inch allowed in the flanges?

Here $W = 40$ tons, $L = 30$ feet, $d = 2$ feet.

$A = $ area of flanges, $s = 6$ tons.

Then at centre,

$$M = \frac{40 \times 30}{4} = 300 \text{ ton-feet};$$

$$A = \frac{300}{2 \times 6} = 25 \text{ square inches.}$$

Let $r = 10$ feet from R' support, and $u = 5$ feet from the same support.

$$\text{At } r, M = \frac{40}{2} \times 10 = 200 \text{ ton-feet,}$$

$$A = \frac{200}{2 \times 6} = 16.66 \text{ square inches.}$$

EXAMPLE I.

At u, $M = \dfrac{40}{2} \times 5 = 100$ ton-feet,

$$A = \dfrac{100}{2 \times 6} = 8.33 \text{ square inches.}$$

Then we require in the flanges for the above girder:

At centre, 25.0 square inches.
5 ft. from " 16.66 " "
10 ft. " " 8.33 " "

Construction of Flanges.—In the results of the example just given, it will be observed that the area of metal required in the flanges increases gradually from the points of support towards the centre of the girder. This will be accomplished by building up the plates of metal overlapping each other for the computed amount.

To make up the 25 square inches in the top flange at the centre, we extend the angles from end to end, and making the girder 12 inches wide, using ordinary size plates (none less than $\frac{3}{8}$ of an inch thick) and angles with the longer leg horizontal, we would require:

Top flange = 2 angles* $5'' \times 4'' \times \frac{1}{2}'' =$ 8.50 square inches.
          1 plate $12'' \times \frac{1}{2}''$ = 6.00 " "
          1 " $12'' \times \frac{1}{2}''$ = 6.00 " "
          1 " $12'' \times \frac{3}{8}''$ = 4.50 " "
                         Total, 25.09 " "

For the bottom flange, the rivet-holes must be deducted to obtain the net section. By referring to the section of the constructed girder, Fig. 19, it will be noticed that the greatest loss

---

* For areas of angles in inches see Table, page 33.

of section is two rivet-holes opposite each other, connecting the angles with the plates of the bottom flange.

Using ¾-inch-diameter rivets, and allowing ⅛ of an inch more for any injury to the metal in the process of punching, we have the area of a rivet-hole equal to $\frac{1}{2}'' + \frac{1}{2}'' + \frac{1}{2}'' + \frac{3}{8}''$ × $1'' = 1\frac{7}{8}$ square inches, for two rivet-holes $2 \times 1\frac{7}{8}'' = 3\frac{3}{4}$ square inches, to be added to the bottom flange, or $25'' + 3\frac{3}{4}'' = 28\frac{3}{4}$ square inches. Then

Bottom flange = 2 angles $5'' \times 4'' \times \frac{1}{2}''$ = 8.50 square inches.
          1 plate $12'' \times \frac{5}{8}''$ = 7.50 " "
          1 " $12'' \times \frac{5}{8}''$ = 7.50 " "
          1 " $12'' \times \frac{7}{16}''$ = 5.25 " "
                             Total, 28.75 " "

Flanges reduced in Area towards the Supports.—To reduce the area of the flanges as the ends are approached, draw the diagram Fig. 15, making R and R' equal to the span of 30

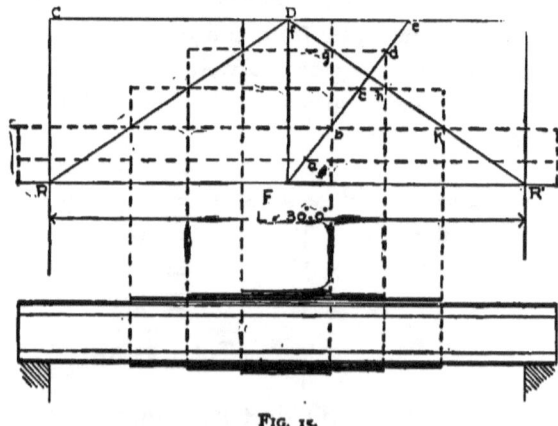

FIG. 15.

feet, and set off FD at centre of span equal to the bending moment at that point, or equal to DF, Fig. 14. Connect RD and DR'. Draw the rectangle $RCER'$.

EXAMPLE I.

Then from F place any scale at any angle, as Fe, until it measures 25 square inches, the number required in the flanges at the centre. For two angles $5'' \times 4'' \times \frac{3}{8}''$, set off 4.25 square inches each at a and b; one plate $12'' \times \frac{1}{2}'' = 6$ square inches at c; one plate $12'' \times \frac{1}{2}'' = 6$ square inches at d; and one plate $12'' \times \frac{3}{8}'' = 4.5$ square inches at e. Horizontal lines drawn from a, b, c, d, and e to DR' at f, g, h, and k, and by the set-square carried down to the base line RR', give the ends of plates in the flanges.

Those flange plates which do not extend from end to end of girder should be run as many inches beyond the point where, according to the calculation we have made, the sectional area of that plate must form the required section, so as to catch sufficient number of rivets in the flange in order to transmit the amount of stress which the plate is required to sustain. Thus for the top plate $12'' \times \frac{3}{8}''$, the amount of stress which it is expected to receive is equal to $12'' \times \frac{3}{8}'' \times 12{,}000 = 54{,}000$ pounds. Now if we have two rows of rivets on the flange, and supposing that each rivet has a safe shearing strain of 4510 pounds per square inch, the plate must be extended to take in at least $\frac{5400}{4510}$, say 12 rivets, beyond the point at which it is calculated to form the flange section. By referring to the diagram it will be seen that the full width of the second plate extends to dh, and we only require the area of same to decrease gradually from g to h, and in like manner in the top plate from f to g. We will therefore be able to place the majority of these 12 rivets of the top plate between f and g.

In all the following examples the plates are extended 12 inches beyond the position determined by the diagram. Plates over $\frac{1}{2}''$ thick should be extended as described above to a sufficient length to prevent the crowding of rivets.

The plates in the bottom flange are practically of the same area as those of the top flange, the loss of area by rivet-holes being made up by the thicker plates; they will therefore be the same length as those in the top flange.

Webs.—The downward pressure at the middle is equal to the upward pressure or reactions at the ends; and since the load is central between the points of support, the reactions of these points are equal, and each is equal to one half the load at centre, which gives 20 tons or 40,000 pounds shear on the web at each support.

$$\text{Then } T = \frac{S}{DK} = \frac{40000}{24 \times 6000} = .28, \text{ or nearly } \tfrac{5}{16} \text{ inch.}$$

The least practicable thickness allowed in the webs of al girders is $\tfrac{3}{8}$ of an inch; we will therefore require in this a $\tfrac{3}{8}''\times$ 24" web plate.

Stiffeners.—To determine whether we require stiffeners, we should first determine whether the web, as already proportioned for shearing, is able to sustain the same compressive stress to resist *buckling*. This condition is attained when the shear per square inch of cross-section at any point does not exceed the

$$\textit{Safe resistance to buckling per square inch} = \frac{10000}{1 + \dfrac{d^2}{3000 t^2}}.$$

When $d =$ depth, and $t =$ thickness of web in inches.

$$\text{Then } \frac{10000}{1 + \dfrac{24 \times 24}{3000 \times \tfrac{3}{8} \times \tfrac{3}{8}}} = 4224 \text{ pounds.}$$

But as we have adopted 6000 pounds per square inch for safe shearing, the web will require to be stiffened throughout at a distance of 3 feet, the shearing strain on the web being

EXAMPLE I.

uniform throughout on a girder with load in centre (see graphical representation of shearing forces, Fig. 19).

Then to stiffen the web at these points, rivet to each side of the web a $4'' \times 4'' \times \frac{3}{8}''$ angle. We then have for the thickness $\frac{3}{8}'' + \frac{3}{8}'' + \frac{3}{8}'' = \frac{9}{8}$ inches, and the formula becomes:

Safe resistance to buckling

$$= \frac{10000}{1 + \dfrac{24 \times 24}{3000 \times \frac{9}{8} \times \frac{9}{8}}} = 8873 \text{ pounds per square inch.}$$

Stiffeners must always be tightly fitted between the flange-angles. In order to bring the stiffeners in contact with web and vertical leg of angle, they are bent as shown in Fig. 16, or fillers may be used as Fig. 17.

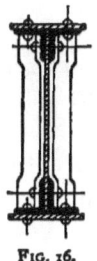
Fig. 16.

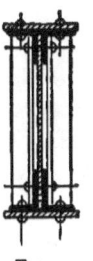
Fig. 17.

The first method requires less material than the second, but requires the work of bending the angles, for which particular dies must be had to give the required amount of bending, and no doubt is more economical where there are a large number of stiffeners to be bent to the same form.

The second method is therefore more preferable for cases when girders to be made are few in number.

Rivets.—Now as to value of rivets through web, we should have for shearing, area of rivet is $\frac{7}{8}'' \times \frac{7}{8}'' \times .7854 = .6013''$ $\times 7500 = 4510$ pounds; being in double shear, twice the value or 9020 pounds.

For bearing $\frac{7}{8}'' \times \frac{3}{8}'' \times 15000 = 4920$ pounds. As the bearing value is the smaller, we will use that in determining the number.

The bending moment at the centre of girder is 300 ton-feet This divided by the depth will give the horizontal flange strain at centre of girder.

$$\text{Then } \frac{300}{2} = 150 \text{ tons or } 300{,}000 \text{ pounds.}$$

This again divided by the least value of a web rivet, which we found to be 4920 pounds, gives the total number of rivets required; or

$$\frac{300000}{4920} = 60 \text{ rivets}$$

in a distance of 180 inches, spaced 3 inches centres.

Graphical Representation of Bending Moments and Shearing Forces.—The bending moments and shearing forces at each point of a girder may be represented graphically by lines laid off to scale, as will be shown by example.

This method will be found far more preferable, on account of the rapidity with which the work can be accomplished, especially for girders with two or more concentrated loads, also with concentrated loads and a uniform load combined. According to our example, we have a girder 30 feet span, 2 feet in depth, to sustain at the middle 40 tons.

Set the load W of 40 tons to a reasonable scale off along a line PP'. The line PP' is thus the polygon of the given force, and $P'P$, its closing line, is the resultant. (For these principles refer to works on Graphic Statics.)

Since the reaction of R and R' must be equal, we take the pole distance O in a horizontal through the centre of the force line PP' at H, and draw the radii OP, OP'; then describe the

EXAMPLE I.

funicular polygon *abc* by drawing *ab* parallel to *PO*, terminating in *W* produced, and *bc* parallel to *P'O*. The funicular polygon is now closed by the line *ca*, and a line *OS* is drawn through pole (on *OH* in this case) which is parallel to *ac* and to the girder.

Then any ordinate from *ac*, as x, y, or z, taken to these inclined lines *ab* or *bc*, multiplied by the pole distance *OH*, will

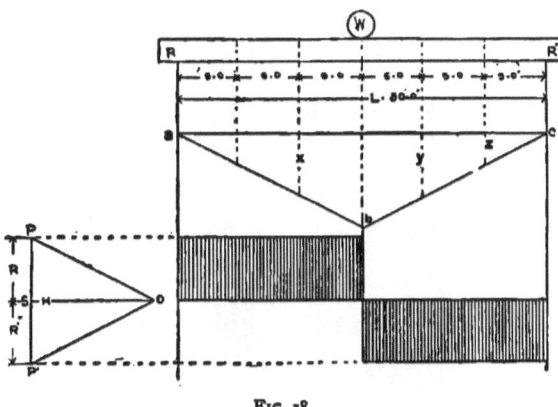

Fig. 18.

give the bending moment at any point of the girder. The pole *O* may be placed at any distance or direction; the result will be the same; but to facilitate calculations, take ten units of the scale adopted.

The shearing forces are equal to the distances *R* and *R'* of the points of the polygon of forces from *S*.

Accordingly, the shearing forces have been taken from the polygon of forces and used as ordinates of the segments of *RR'* to which they correspond. Thus the hatched figure is obtained, which is termed the *shearing force* diagram, and the vertical ordinates of this diagram give the shearing force at any section of the girder *RR'*, and in this particular example it can be seen at a glance and, as previously found, that the shear on the web is uniform from centre to ends.

COMPOUND RIVETED GIRDERS.

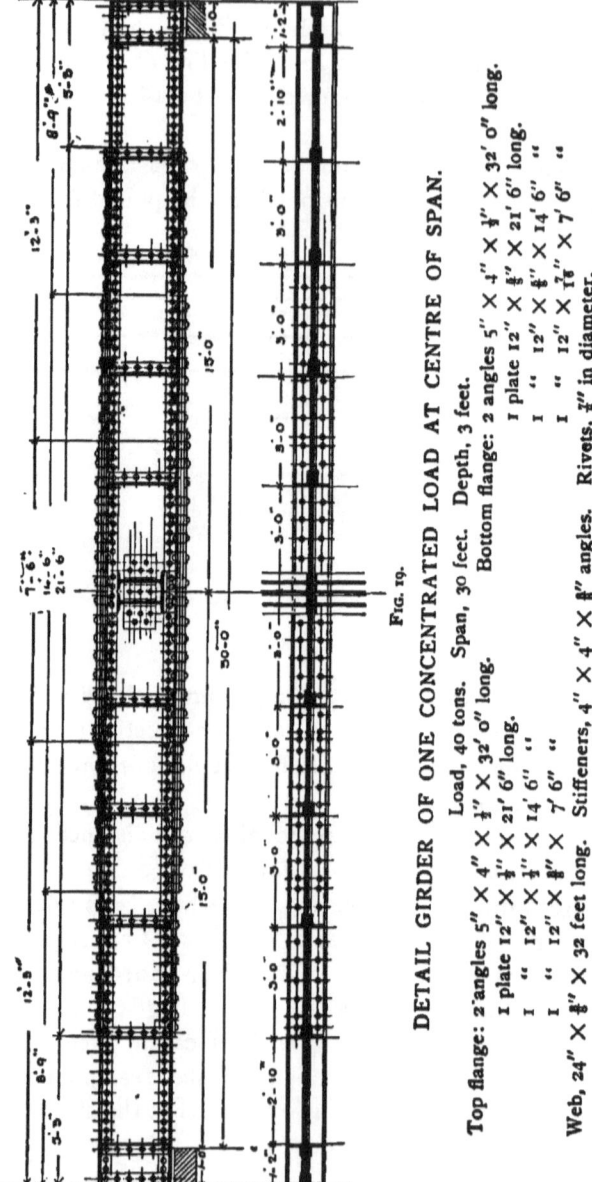

Fig. 19.

DETAIL GIRDER OF ONE CONCENTRATED LOAD AT CENTRE OF SPAN.

Load, 40 tons.   Span, 30 feet.   Depth, 3 feet.

Top flange: 2 angles 5″ × 4″ × ½″ × 32′ 0″ long. Bottom flange: 2 angles 5″ × 4″ × ½″ × 32′ 0″ long.
1 plate 12″ × ½″ × 21′ 6″ long.   1 plate 12″ × ⅝″ × 21′ 6″ long.
1 "   12″ × ½″ × 14′ 6″ "   1 "   12″ × ⅝″ × 14′ 6″ "
1 "   12″ × ⅜″ × 7′ 6″ "   1 "   12″ × ⁷⁄₁₆″ × 7′ 6″ "

Web, 24″ × ⅜″ × 32 feet long.   Stiffeners, 4″ × 4″ × ⅜″ angles.   Rivets, ⅞″ in diameter.

EXAMPLE I.

AREAS OF ANGLES

WITH EVEN LEGS.

Size in Inches.	Thickness in Inches.											
	$\frac{1}{4}$	$\frac{5}{16}$	$\frac{3}{8}$	$\frac{7}{16}$	$\frac{1}{2}$	$\frac{9}{16}$	$\frac{5}{8}$	$\frac{11}{16}$	$\frac{3}{4}$	$\frac{13}{16}$	$\frac{7}{8}$	1
6 × 6	....	....	....	5.06	5.75	6.43	7.11	7.78	8.44	9.06	9.74	11.0
5 × 5	....	....	....	4.18	4.75	5.31	5.86	6.42	6.94	7.47	7.99	9.0
4 × 4	....	....	2.86	3.31	3.75	4.18	4.61	5.03	5.44	5.84		
3½ × 3½	....	....	2.48	2.87	3.25	3.62	3.99	4.34	4.69	5.03		
3 × 3	1.44	1.78	2.11	2.43	2.75	3.06	3.36	3.65				
2¾ × 2¾	1.31	1.62	1.92	2.21	2.50							
2½ × 2½	1.19	1.47	1.73	2.00	2.25							
2¼ × 2¼	1.06	1.31	1.55	1.78	2.00							
2 × 2	0.94	1.15	1.36	1.56								
1¾ × 1¾	0.81	1.00	1.17	1.30								
1½ × 1½	0.69	0.84	0.99									

WITH UNEVEN LEGS.

Size	$\frac{1}{4}$	$\frac{5}{16}$	$\frac{3}{8}$	$\frac{7}{16}$	$\frac{1}{2}$	$\frac{9}{16}$	$\frac{5}{8}$	$\frac{11}{16}$	$\frac{3}{4}$	$\frac{13}{16}$	$\frac{7}{8}$	1
7 × 3½	....	....	....	4.40	5.00	5.59	6.17	6.75	7.31	7.87	8.42	9.50
6 × 4	....	....	3.61	4.18	4.75	5.30	5.86	6.41	6.94	7.47	7.99	9.00
6 × 3½	....	....	3.42	3.96	4.50	5.03	5.55	6.06	6.56	7.06	7.55	8.50
5 × 4	....	....	3.23	3.74	4.25	4.74	5.23	5.72	6.18	6.65	7.11	8.00
5 × 3½	....	....	3.05	3.52	4.00	4.46	4.92	5.37	5.81	6.25	6.67	
5 × 3	....	....	2.86	3.30	3.75	4.17	4.60	5.03	5.44	5.84		
4½ × 3	....	....	2.67	3.09	3.50	3.90	4.30	4.68	5.06	5.43		
4 × 3½	....	....	2.67	3.09	3.50	3.90	4.30	4.68	5.06	5.43		
4 × 3	....	2.09	2.48	2.87	3.25	3.62	3.98	4.34	4.69	5.03		
3½ × 3	....	1.93	2.30	2.65	3.00	3.34	3.67	4.00	4.31	4.62		
3½ × 2½	1.44	1.78	2.11	2.43	2.75	3.06	3.36	3.65				
3 × 2½	1.31	1.62	1.92	2.21	2.50	2.78						
3 × 2	1.19	1.46	1.73	1.99	2.25							
2½ × 2	1.06	1.31	1.55	1.78	2.00							
2½ × 1½	0.88	1.07	1.27	1.45	1.63							
2 × 1½	0.78											

SECTIONAL AREA IN INCHES OF RIVET-HOLES IN PLATES OF VARIOUS THICKNESSES, TAKEN ⅛ INCH IN EXCESS OF DIAMETER OF RIVET.

Thickness of Plate.	Number of Rivets 1 inch Diameter.								Number of Rivets ⅞ inch Diameter.							
	1	2	3	4	5	6	7	8	1	2	3	4	5	6	7	8
¼	1.12	2.25	3.37	4.50	5.62	6.75	7.87	9.00	1.00	2.00	3.00	4.00	5.00	6.00	7.00	8.00
	1.05	2.10	3.16	4.21	5.27	6.32	7.38	8.43	0.94	1.87	2.81	3.75	4.68	5.62	6.56	7.50
	.98	1.97	2.95	3.93	4.92	5.90	6.89	7.87	.87	1.75	2.62	3.50	4.37	5.25	6.12	7.00
	.91	1.83	2.74	3.65	4.57	5.48	6.39	7.31	.81	1.62	2.44	3.25	4.06	4.87	5.68	6.50
	.84	1.69	2.53	3.37	4.22	5.06	5.90	6.75	.75	1.50	2.25	3.00	3.75	4.50	5.25	6.00
	.77	1.55	2.32	3.09	3.86	4.64	5.41	6.19	.69	1.37	2.06	2.75	3.43	4.12	4.81	5.50
	.70	1.41	2.11	2.81	3.51	4.22	4.92	5.62	.62	1.25	1.87	2.50	3.12	3.75	4.37	5.00
	.63	1.26	1.90	2.53	3.16	3.80	4.42	5.06	.56	1.12	1.69	2.25	2.81	3.37	3.93	4.50
	.56	1.11	1.69	2.25	2.81	3.37	3.94	4.50	.50	1.00	1.50	2.00	2.50	3.00	3.50	4.00
1/16	.49	.98	1.47	1.97	2.46	2.95	3.44	3.94	.44	.87	1.31	1.75	2.18	2.62	3.06	3.50
	.42	.84	1.26	1.69	2.11	2.53	2.95	3.37	.37	.75	1.12	1.50	1.87	2.25	2.62	3.00
	.35	.70	1.05	1.40	1.76	2.11	2.46	2.81	.31	.62	.94	1.25	1.56	1.87	2.18	2.50
	.28	.56	.84	1.12	1.40	1.69	1.97	2.25	.25	.50	.75	1.00	1.25	1.50	1.75	2.00

Thickness of Plate.	Number of Rivets ¾ inch Diameter.								Number of Rivets ⅝ inch Diameter.							
	1	2	3	4	5	6	7	8	1	2	3	4	5	6	7	8
1	0.87	1.75	2.62	3.50	4.37	5.25	6.12	7.00	0.75	1.50	2.25	3.00	3.75	4.50	5.25	6.00
	.82	1.64	2.46	3.28	4.10	4.92	5.74	6.56	.70	1.40	2.11	2.81	3.51	4.22	4.92	5.62
	.77	1.53	2.30	3.06	3.83	4.59	5.36	6.12	.65	1.31	1.96	2.62	3.28	3.94	4.59	5.25
	.71	1.42	2.13	2.84	3.55	4.26	4.97	5.68	.61	1.22	1.83	2.44	3.04	3.65	4.26	4.87
	.66	1.31	1.96	2.62	3.28	3.93	4.59	5.25	.56	1.12	1.69	2.25	2.81	3.37	3.93	4.50
	.60	1.20	1.80	2.40	3.00	3.60	4.21	4.81	.51	1.03	1.54	2.06	2.57	3.09	3.60	4.12
	.55	1.09	1.64	2.19	2.73	3.28	3.83	4.38	.47	.94	1.41	1.88	2.34	2.81	3.28	3.75
	.49	.98	1.48	1.96	2.46	2.95	3.44	3.94	.42	.84	1.26	1.69	2.10	2.53	2.95	3.37
	.43	.87	1.31	1.75	2.18	2.62	3.06	3.50	.37	.75	1.12	1.50	1.87	2.25	2.62	3.00
1/16	.38	.76	1.15	1.53	1.91	2.30	2.68	3.06	.33	.66	.98	1.31	1.64	1.07	2.29	2.62
	.32	.65	.98	1.31	1.64	1.97	2.30	2.62	.28	.56	.84	1.12	1.40	1.69	1.97	2.25
	.27	.55	.82	1.09	1.36	1.64	1.91	2.18	.23	.47	.70	.94	1.17	1.40	1.64	1.87
	.22	.44	.66	.87	1.09	1.31	1.53	1.75	.18	.37	.56	.75	.93	1.12	1.31	1.50

EXAMPLE I.

TABLE SHOWING GROSS AREA OF PLATES OF VARIOUS THICKNESSES.

Thickness of Plate.	Width of Plate in Inches.													
	8	9	10	12	14	15	16	18	20	21	22	24	26	28	30
1	8.00	9.00	10.00	12.00	14.00	15.00	16.00	18.00	20.00	21.00	22.00	24.00	26.00	28.00	30.00
15/16	7.50	8.43	9.37	11.25	13.12	14.06	15.00	16.87	18.75	19.68	20.62	22.50	24.37	26.25	28.12
7/8	7.00	7.87	8.75	10.50	12.25	13.12	14.00	15.75	17.50	18.37	19.25	21.00	22.75	24.50	26.25
13/16	6.50	7.31	8.12	9.75	11.37	12.18	13.00	14.62	16.25	17.06	17.87	19.50	21.12	22.75	24.37
3/4	6.00	6.75	7.50	9.00	10.50	11.25	12.00	13.50	15.00	15.75	16.50	18.00	19.50	21.00	22.50
11/16	5.50	6.18	6.87	8.25	9.62	10.31	11.00	12.37	13.75	14.43	15.12	16.50	17.86	19.25	20.62
5/8	5.00	5.62	6.25	7.50	8.75	9.37	10.00	11.25	12.50	13.12	13.75	15.00	16.25	17.50	18.75
9/16	4.50	5.06	5.62	6.75	7.87	8.43	9.00	10.12	11.25	11.81	12.37	13.50	14.62	15.75	16.87
1/2	4.00	4.50	5.00	6.00	7.00	7.50	8.00	9.00	10.00	10.50	11.00	12.00	13.00	14.00	15.00
7/16	3.50	3.94	4.37	5.25	6.12	6.56	7.00	7.87	8.75	9 19	9.62	10.50	11.37	12.25	13.12
3/8	3.00	3.37	3.75	4.50	5.25	5.62	6 00	6.75	7.50	7.87	8.25	9.00	9 75	10.50	11.25
5/16	2.50	2.81	3.12	3.75	4.37	4.69	5.00	5.62	6.25	6.56	6.87	7.50	8.12	8.75	9.37
1/4	2.00	2.25	2.50	3.00	3 50	3.75	4.00	4.50	5 00	5.25	5.50	6.00	6.50	7.00	7.50

EXPLANATION.—Required the sectional area of a plate 26″ × 11/16″ punched by six 7/8″ rivets. The gross area by table = 17.86 square inches, and the area of six 7/8″ rivets by the previous table = 3.60; that is, 17.86 − 3.60 = 14.26, the area required.

Required the sectional area of a 6″ × 4″ × 7/8″ angle punched by two 7/8″ rivets. The gross area by table of "Areas of Angles" = 7.99 square inches, and the area of two 7/8″ rivets through a 7/8″ plate by the table = 1.53. Then 7.99 − 1.53 = 6.46, the area required.

SAFE BUCKLING VALUE OF WEB PLATES PER SQUARE INCH (WROUGHT-IRON).

Calculated by formula $\dfrac{10000}{1 + \dfrac{d^2}{3000 t^2}}$.

d = depth in inches.
t = thickness in inches.

Thickness in Inches.	Depth in Inches.										
	20	24	28	30	32	36	40	42	48	50	52	60
1/4	3195	2455										
5/16	4220	3365	2714									
3/8	5134	4229	3498	3192	2889	2456	2087	1930	1548	1442	1350	
7/16	5900	4992	4228	3896	3624	3069	2696	2455	1994	1868	1751	
1/2	6522	5652	4890	4546	4228	3666	3191	2983	2543	2308	2172	1724
9/16	7035	6223	5476	5133	4787	4229	3724	3498	2918	2749	2599	2087
5/8	7456	6704	5932	5656	5339	4748	4228	3992	3371	3191	3024	2456
11/16	7818	7133	6461	6143	5834	5252	4726	4500	3835	3645	3465	2848
3/4	8044	7612	6828	6522	6226	5656	5133	4889	4228	4030	3885	3191
13/16		7752	7164	6868	6585	6045	5534	5290	4623	4420	4224	3549
7/8				7184	6920	6392	5882	5649	4992	4780	4593	3891
15/16						6700	6211	5988	5336	5025	4926	4237
1									5674	5455	5263	4545

N.B.—If the buckling value is less than the shearing (6000 pounds for wrought-iron), the web will require to be stiffened.

SHEARING VALUE OF WEB PLATES, WROUGHT-IRON, 6000 LBS. PER SQUARE INCH.

Depth in Inches	Thickness of Plate.								
	$\frac{3}{8}$	$\frac{7}{16}$	$\frac{1}{2}$	$\frac{9}{16}$	$\frac{5}{8}$	$\frac{11}{16}$	$\frac{3}{4}$	$\frac{7}{8}$	1
12	27000	31500	36000	40500	45000	49500	54000	63000	72000
14	31500	36750	42000	47250	52500	57750	63000	73500	84000
16	36000	42000	48000	54000	60000	66000	72000	84000	96000
18	40500	47250	54000	60750	67500	74250	80500	94500	108000
20	45000	52500	60000	67500	75000	82500	90000	105000	120000
22	49500	57750	66000	74250	82500	90750	99000	115500	132000
24	54000	63000	72000	81000	90000	99000	108000	126000	144000
26	58500	68250	78000	87750	97500	107250	117000	136500	156000
28	63000	73500	84000	94500	105000	115500	126000	147000	168000
30	67500	78750	90000	101250	112500	123750	135000	157500	180000
32	72000	84000	96000	108000	120000	132000	144000	168000	192000
34	76500	89250	102000	114750	127500	140250	153000	178500	204000
36	81000	94500	108000	121500	135000	148500	162000	189000	216000
38	85500	99750	114000	128250	142500	156750	171000	199500	228000
40	90000	105000	120000	135000	150000	165000	180000	210000	240000
42	94500	110250	126000	141750	157500	173250	189000	220500	252000
44	99000	115500	132000	148500	165000	181500	198000	231000	264000
46	103500	120750	138000	155250	172500	189750	207000	241500	278000
48	108000	126000	144000	162000	180000	198000	216000	252000	288000
50	112500	131250	150000	168750	187500	206250	225000	262500	300000
52	117000	136500	156000	175500	195000	214500	234000	273000	312000
54	121500	141750	162000	182250	202500	222750	243000	283500	324000
56	126000	147000	168000	189000	210000	231000	252000	294000	336000
58	130500	152250	174000	195750	217500	239250	261000	304500	348000
60	135000	157500	180000	202500	225000	247500	270000	315000	360000

EXAMPLE I.

SHEARING VALUE OF WEB PLATES, WROUGHT-STEEL, 7000 LBS. PER SQUARE INCH.

Depth in Inches	Thickness of Plate.								
	$\frac{3}{8}$	$\frac{7}{16}$	$\frac{1}{2}$	$\frac{9}{16}$	$\frac{5}{8}$	$\frac{11}{16}$	$\frac{3}{4}$	$\frac{7}{8}$	1
12	31500	36750	42000	47250	52500	57750	63000	73500	84000
14	36750	42875	49000	55125	61250	67375	73500	85750	98000
16	42000	49000	56000	63000	70000	77000	84000	98000	112000
18	47250	55125	63000	70875	78750	86625	94500	110250	126000
20	52500	61250	70000	78750	87500	96250	105000	122500	140000
22	57750	67375	77000	86625	96250	105875	115500	134750	154000
24	63000	73500	84000	94500	105000	115500	126000	147000	168000
26	68250	79625	91000	102375	113750	125125	136500	159250	182000
28	73500	85750	98000	110250	122500	134750	147000	171500	196000
30	78750	91875	105000	118125	131250	144375	157500	183750	210000
32	84000	98000	112000	126000	140000	154000	168000	196000	224000
34	89250	104125	119000	133875	148750	163625	178500	208250	238000
36	94500	110250	126000	141750	157500	173250	189000	220500	252000
38	99750	116375	133000	149625	166250	182875	199500	232750	266000
40	105000	122500	140000	157500	175000	192500	210000	245000	280000
42	110250	128625	147000	165375	183750	202125	220500	257250	294000
44	115500	134750	154000	173250	192500	211750	231000	269500	308000
46	120750	140875	161000	181125	201250	221375	241500	281750	322000
48	126000	147000	168000	189000	210000	231000	252000	294000	336000
50	131250	153125	175000	196875	218750	240625	262500	306250	350000
52	136500	159250	182000	204750	227500	250250	273000	318500	364000
54	141750	165375	189000	212625	236250	259875	283500	330750	378000
56	147000	171500	196000	220500	245000	269500	294000	343000	392000
58	152250	177625	203000	228375	253750	279125	304500	355250	406000
60	157500	183750	210000	236250	262500	288750	315000	367500	420000

Example II.

GIRDER SUPPORTING A CONCENTRATED LOAD NOT AT CENTRE OF SPAN.

In a girder supported at both ends with a load concentrated *not* at centre of span, the maximum bending moment is at the load, and is equal to the load multiplied by the distance from load to left support, and from load to right support divided by the span; or,

$$M = W \times \frac{a \times b}{L}.$$

To find the bending moment at any point in the girder when

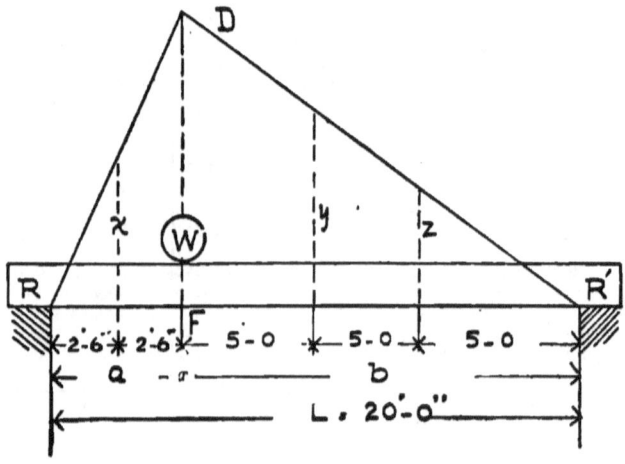

Fig. 20.

the load is *not* at centre, make FD by any scale, Fig. 20, equal M at the load; join RD and DR'. Then x, y, z, by the same scale, will measure the moments at their respective points in the girder.

$$\text{At } x, M = W \times \frac{Rx \times R'F}{L}.$$

$$\text{At } y, M = W \times \frac{RF \times R'y}{L}.$$

$$\text{At } z, \dot{M} = W \times \frac{RF \times R'z}{L}.$$

Example: What metal area would be required in the flanges of a plate girder of 20 feet span, 2 feet 6 inches in depth, to sustain 60 tons concentrated at 5 feet from left support?

Here $W = 60$ tons; $L = 20$ feet; $d = 2$ feet 6 inches; $s = 6$ tons. Then

$$\text{At } F, M = 60 \times \frac{5 \times 15}{20} = 225 \text{ ton-feet,}$$

$$A = \frac{225}{2.5 \times 6} = 15 \text{ square inches.}$$

$$\text{At } x, M = 60 \times \frac{2.5 \times 15}{20} = 112.5 \text{ ton-feet,}$$

$$A = \frac{112.5}{2.5 \times 6} = 7.5 \text{ square inches.}$$

$$\text{At } y, M = 60 \times \frac{5 \times 10}{20} = 150 \text{ ton-feet,}$$

$$A = \frac{150}{2.5 \times 6} = 10 \text{ square inches.}$$

At z, $M = 60 \times \dfrac{5 \times 5}{20} = 75$ ton-feet,

$$A = \dfrac{75}{2.5 \times 6} = 5 \text{ square inches.}$$

Then we require in the flanges for the above girder:

At the load, 15.0 square inches.
At x, 2 feet 6 inches from "   7.50 "   "
At y, 5 "   .   "   "   10.00 "   "
At z, 10 "   "   "   5.00 "   "

Construction of Flanges.—To make up the maximum section under the load we would require:

Top flange = 2 angles $5'' \times 3'' \times \frac{1}{2}'' = $ 7.5 square inches.
            1 plate $12'' \times \frac{5}{8}''$ = 7.5   "   "
                                      Total, 15.0   "   "

For the bottom flange, rivet-holes to be deducted to obtain the net section. The loss of metal by rivet-holes is the same in this as the former example.

Using $\frac{7}{8}$-inch-diameter rivets, and allowing $\frac{1}{8}$ of an inch more for any injury to the metal by punching, area of rivet-hole equals $\frac{5}{8}'' + \frac{1}{2}'' \times 1'' = \frac{9}{8}$ square inches; for two rivet-holes $2'' \times \frac{9}{8}'' = 2.25$ square inches, to be added to the bottom flange, or $15'' + 2.25'' = 17.25$ square inches.

Bottom flange = 2 angles $5'' \times 3'' \times \frac{1}{2}'' = $ 7.5 square inches.
               1 plate $12'' \times 1\frac{3}{16}''$ = 9.75 "   "
                                         Total, 17.25 "   "

EXAMPLE II.

Flanges reduced in Area towards the Supports. — To place the plates in their required position for the calculated area: Draw the diagram, Fig. 21, making R and R' equal to the

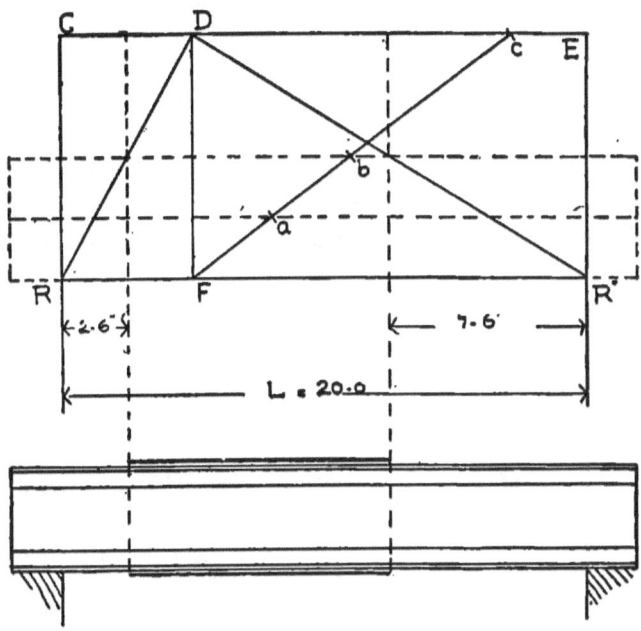

FIG. 21.

span of 20 feet as in the previous example, and set off lines FD at position of load equal to the maximum bending moment at that point, or equal to FD, Fig. 20. Connect RD and DR'. Draw the rectangle $RCER'$. Then from F place any scale at any angle, as Fc, until it measures 15, the number of square inches required in the flanges at F or at the load. For two angles set off 3.75 square inches at a and b, and one plate 12 $\times 1\frac{3}{8}$ inches, or 7.50 square inches, at C. Horizontal lines drawn from a, b, and c to DR, DR', and carried down to base line RR' give the position of the plates. The angles to extend from end to end as shown.

Both top and bottom plates to extend 12 inches each way from *FD*, in addition to that determined by the diagram.

Webs.—The bearing area upon which a girder is supported reacts against the girder an amount equal to the pressure of the load upon them; or, the sum of the loads on the girder is equal to the sum of the reactions. Hence, if there be but one support as in a cantilever, this condition gives at once the reaction. For a uniform load and a concentrated load at centre, on two supports, it is evident that each reaction equals one half of the load.

We have in this example a single concentrated load situated at 5 feet from left support, whose span is 20 feet.

$$\text{Then } R \text{ supports } \frac{60 \times 15}{20} = 45 \text{ tons,}$$

$$\text{and } R' \text{ `` } \frac{60 \times 5}{20} = 15 \text{ tons.}$$

(Refer to article Shearing Forces on the Webs.)

The thickness of the web may then be determined as in the previous example.

$$\text{At } R, \ T = \frac{s}{dk} = \frac{90,000}{30 \times 6000} = \tfrac{1}{2} \text{ inch.}$$

$$\text{At } R', \ T = \frac{30,000}{30 \times 6000} = \tfrac{1}{6} \text{ inch.}$$

The one-half-inch thickness to be adopted and used from end to end of girder; otherwise we would require a variety of thicknesses to be made up between each end, which is altogether impracticable in small girders.

EXAMPLE II.

Stiffeners.—To determine whether we require stiffeners:

Safe resistance to buckling

$$= \frac{10000}{1 + \frac{d^2}{3000 t^2}} = \frac{10000}{1 + \frac{30 \times 30}{3000 \times \frac{1}{2} \times \frac{1}{2}}} = 4590 \text{ lbs. per sq. inch.}$$

But as we have adopted 6000 pounds for safe shearing, the web will have to be stiffened throughout, at the bearings, under the load, and every 3 feet, by $4'' \times 4'' \times \frac{3}{8}''$ angles riveted each side of web. We have then for the thickness $\frac{3}{8}'' + \frac{1}{2}'' + \frac{3}{8}''$ = $\frac{10}{8}$ inches, and the formula becomes

$$\textit{Safe resistance to buckling} = \frac{10,000}{1 + \frac{30 \times 30}{3000 \times \frac{10}{8} \times \frac{10}{8}}} = 8389 \text{ lbs.}$$

Rivets.—This example being a single-web girder, the rivets will be in double shear; so we will take their bearing value, which is less than the shearing. The bearing area $= \frac{7}{8}'' \times \frac{1}{2}''$ $\times 15000 = 6562$ pounds. The bending moment at load is 225 ton-feet; this divided by the depth gives the horizontal flange strain each way from position of load to end of girder.

$$\text{Then } \frac{225}{2.5} = 90 \text{ tons or } 180,000 \text{ pounds.}$$

This again divided by the value of the rivet gives the total number of rivets required; or,

$$\frac{180,000}{6562} = 27 \text{ rivets,}$$

in a distance of 60 inches from the load to R support spaced about $2\frac{1}{4}$ inches centres.

This is *closer* than they should be spaced in a straight line, so we will have to stagger them as much as possible. (See girder drawing, Fig. 23.) We will also require 27 rivets from the position of load to R', a distance of 180 inches, spaced 6.6 inches. This is *more* than they should be spaced in a straight line; as our maximum is 6 inches, we will space them accordingly. Had we used the shearing stress for the rivet spacing, we would have a uniform shear of 90,000 pounds from position of load to R, and a uniform shear of 30,000 pounds to R'.

At R, $\dfrac{90,000}{2.5 \times 12} = 3000$ pounds per inch of run.

Spaced, $\dfrac{6562}{3000} = 2.19$ inches centre to centre.

At R', $\dfrac{30,000}{2.5 \times 12} = 1000$ pounds per inch of run;

Spaced, $\dfrac{6562}{1000} = 6.56$ inches centre to centre.

Graphical Representation of Bending Moments and Shearing Forces in a Girder with One Concentrated Load not at Centre.—According to our example we have a girder of 20 feet span, sustaining a load of 60 tons 5 feet from left support. Set the load W of 60 tons to a reasonable scale off along the line PP', Fig. 22. The line PP' is thus the polygon of the given force, and $P'P$, its closing line, is the resultant.

Take any point O as pole, equal to ten units of the scale adopted, and draw the radii OP and OP'. Then describe the funicular polygon abc by drawing ab parallel to OP, terminating in W produced, and bc parallel to OP', terminating in the prolongation downwards of R'. The funicular polygon is

EXAMPLE II.

now closed by the line *ca*, and a line OS is drawn through pole O parallel to *ac*. The bending moments are then found in the same manner as described under Fig. 18. Then the re-

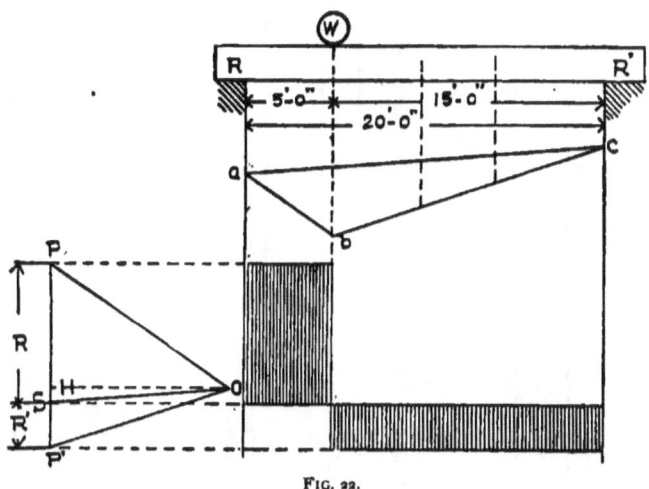

FIG. 22.

action of R will equal the distance by same scale from S to P, and the reaction at R the distance from S to P'.

Or as a condition of equilibrium,

$$\text{the reaction at } R = PS, \text{ and at } R' = SP'.$$

The shearing forces, as in the previous example, are taken from the hatched figure. By referring to the diagram it will be noticed that the greatest shear on the web is at R support measured from P to S, and only a small percentage of the load W is sustained by R' support.

46 COMPOUND RIVETED GIRDERS.

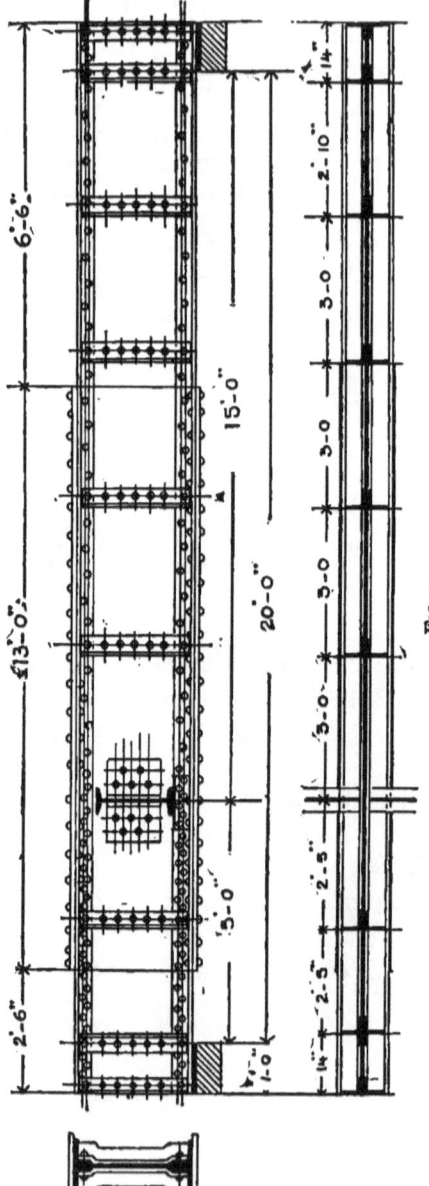

Fig. 23.

DETAIL GIRDER SUPPORTING A CONCENTRATED LOAD NOT AT CENTRE OF SPAN.

Load, 60 tons. Span, 20 feet. Depth, 2 feet 6 inches.

Top flange: 2 angles 5" × 3" × ½" × 22' 0" long.   Bottom flange: 2 angles 5" × 3½" × ½" × 22' 0" long.
1 plate 12" × ⅜" × 13' 0" long.   1 plate 12" × 11/16" × 13' 0" long.

Web, 30" × ⅜" × 22' 0". Stiffeners, 4" × 4" × ⅜" angles. Rivets, ⅞" diameter.

Example III.

GIRDER SUPPORTING A UNIFORMLY DISTRIBUTED LOAD.

In a girder supported at both ends with a load distributed over its entire length, the maximum bending moment is at the centre, and is equal to one half the load multiplied by one quarter the span, or

$$M = \frac{WL}{8}.$$

To find the bending moment at any point in the girder, when the load is uniformly distributed:

Make *FD* by scale, Fig. 24, equal *M* at centre of span, and draw the parabola *RDR'* (see method of drawing parabolas,

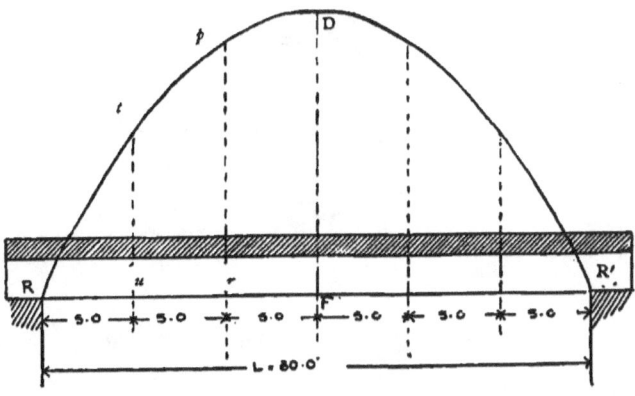

FIG. 24.

Figs. 26 and 27). Then *rp* measured by the same scale will equal the bending moment at point *r*, and *ut* will equal the moment at point *u*.

48 COMPOUND RIVETED GIRDERS.

Or the moment at any point $r = $ half load on $rR \times rR'$, or half load on $rR' \times rR$.

$$\text{At } r, M = \frac{\frac{W}{L} \times rR}{2} \times rR'.$$

$$\text{At } u, M = \frac{\frac{W}{L} \times uR}{2} \times uR'.$$

Example: What metal area would be required in the flanges at the centre of a *box girder* of 30 feet span, 3 feet in depth, to sustain 200 tons of a 16-inch wall distributed over its entire length.

Here $W = $ 200 tons, $L = $ 30 feet, $d = $ 3 feet,
$A = $ area of flange, $s = $ 6 tons.

$$\text{At centre, } M = \frac{200 \times 30}{8} = 750 \text{ ton-feet},$$

$$A = \frac{750}{3 \times 6} = 41.66 \text{ square inches}.$$

$$\text{At } r, M = \frac{\frac{200}{30} \times 10}{2} \times 20 = 666.66 \text{ ton-feet},$$

$$A = \frac{666.66}{3 \times 6} = 37.04 \text{ square inches}.$$

$$\text{At } u, M = \frac{\frac{200}{30} \times 5}{2} \times 20 = 416.66 \text{ ton-feet},$$

$$A = \frac{416.66}{3 \times 6} = 23.15 \text{ square inches}.$$

EXAMPLE III.

Then for the above girder we require in the flanges:

At centre, 41.66 square inches.
5 feet from "   37.04   "   "
10 "   "   "   23.15   "   "

Construction of Flanges.—To make up the maximum section at the maximum bending moment we would require:

Top flange = 2 angles $4'' \times 4'' \times \frac{13}{16}'' = 11.68$ square inches.
3 plates each $16'' \times \frac{5}{8}'' = 30.00$ "   "

Total, 41.69 "   "

For the bottom flange the loss by rivet-holes will be the thickness of plates and one angle, and then using $\frac{7}{8}$-inch-diameter rivets and allowing $\frac{1}{8}$ inch more, we have $\frac{5}{8}'' + \frac{5}{8}'' + \frac{5}{8}'' + \frac{13}{16}'' \times 1 = 2\frac{11}{16}$ square inches, for two rivets $2 \times 2\frac{11}{16}'' = 5.375$ square inches, to be added to the bottom flange at centre, or $41.66'' + 5.375'' = 47$ square inches.

Bottom flange = 2 angles $4'' \times 4'' \times \frac{13}{16}'' = 11.68$ square inches.
3 plates each $16'' \times \frac{3}{4}'' = 36.00$ "   "

Total, 47.68 "   "

Flanges Reduced in Area towards the Supports.—To place the plates of the flanges in their required position for the calculated area, draw the diagram Fig. 25. From F in centre of span, make FD, by the same scale as in Fig. 24, equal to the maximum bending moment at that point. Draw the rectangle $RCER'$. From F place the scale at any angle, as at Fe, until it measures 41.66 or 41.68 square inches.

For two angles set off 5.84 square inches, each, at a and b, and three plates $16'' \times \frac{3}{4}''$, or 12 square inches, each, at c, d, and e. Horizontal lines drawn from a, b, c, d, and e to the parabola RDR', and carried down to base line RR', will give the position of the plates in the flanges.

The angles to extend from end to end of girder, and the adjoining plates are required to extend in like manner for practical reasons, which will be readily seen in all box girders.

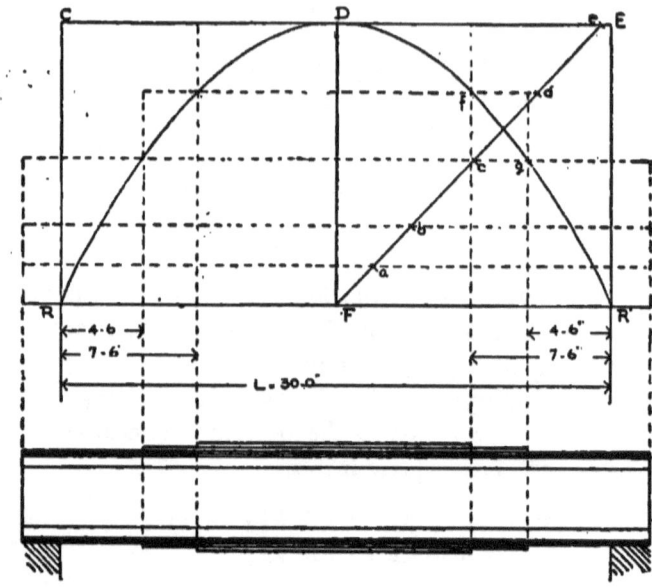

FIG. 25.

The plates of the bottom flange are, for the reasons explained in the previous example, practically the same length as those of the top flange, and should extend 12 inches beyond the calculated length.

Webs.—The reactions on the supports of a girder sustaining a uniformly distributed load are each equal to one half the total load, and the shearing force on the webs at each end of the girder is equal to $\frac{400,000}{2} = 200,000$ pounds.

Then $t = \dfrac{20,0000}{36 \times 6000} = .92$, nearly $1\frac{5}{8}$ of an inch;

EXAMPLE III.

but as we have two webs, each will be equal to one half of $\frac{11}{16}$ or $\frac{11}{32}$, say $\frac{1}{2}$ inch, for the thickness of each web.

Stiffeners.—To determine whether we require stiffeners, only one web need be taken into consideration; and if stiffeners are needed, an angle can be riveted on the outside. Frequently there is considerable shear on the webs; angles are then riveted inside and outside of each web.

Safe resistance to buckling

$$= \frac{10000}{1 + \frac{36 \times 36}{3000 \times \frac{1}{2} \times \frac{1}{2}}} = 3663 \text{ pounds per square inch}$$

The webs will have to be stiffened at the bearings.

At 5 feet from the bearing or supports the shear on the webs is equal to $\frac{2}{3}$ the shear at the bearings, or 133,333 pounds. The safe shear against buckling of a 36-inch web $\frac{1}{2}$ inch thick is 3663 pounds per square inch, as found above. The shearing area of the web at 5 feet from the bearing is $36 \times \frac{1}{2} = 18$ square inches.

Then $3663 \times 18 = 65{,}934$ pounds safe against buckling at that point, and the shear on one web at the same point is $\frac{133{,}333}{2} = 66{,}666$ pounds; a stiffener is therefore required, and one 2′ 6″ towards the bearings, but *none* towards the centre, as the shear is theoretically nothing at the middle of a girder uniformly loaded, but from thence increases by equal increments towards each support (refer to shearing force diagram, Fig. 28).

Rivets.—The rivets connecting the webs to the flanges in a *box girder* are in single shear; therefore the shearing value will be 7500 pounds per square inch, and is measured on the area of the cross-section of the rivet.

The area of a $\frac{7}{8}$-inch-diameter rivet $= \frac{7}{8}'' \times \frac{7}{8}'' \times .7854 = .6013$ (see Table of Shearing and Bearing Resistance of Rivets);

this multiplied by 7500 = 4510, the safe amount of strain each rivet can sustain without shearing.

If the maximum horizontal strain in the flanges is divided by 4510, there results the minimum number of rivets required either way from the centre. Then from the example:

Maximum $M = 750$ ton-feet;

Horizontal strain $= \dfrac{750}{3} = 250$ tons or 500,000 pounds;

and divided by 4510 = 110 rivets, to be placed a distance of 180 inches for one side, 360 inches for both, spaced about 3 inches centres.

On account of the horizontal increments of strain in the web increasing towards the ends, the rivets should be spaced closer as the ends are approached, say 2⅝ inches for the first 5 feet, 3 inches for the next, and 4 inches for the remaining distance.

Method of Drawing Parabolas.—Draw a horizontal RF, Fig. 26, equal to half span of girder. Set off DF perpendicular to RF, making the former equal by scale to the bending moment at that point. Through D draw CD parallel and equal to RF. The ordinates from any points in CD to the parabola will be proportional to the square of the distances of those points from D. Thus if the ordinate at a be 1, then the ordinate at b, twice the distance of a from D, must be 4; and so on.

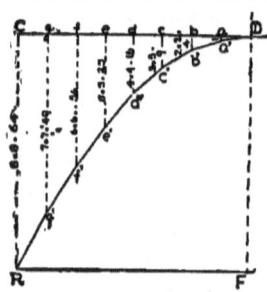

Fig. 26.

To proceed practically, divide CD into a number of equal parts (n) as at a, b, c, etc. Then if RC be divided into (n^2) parts, each of these parts will be the required unit, one of which is the offset at a; four at b; nine at c; and so on. Through the points. a', b', c', etc., thus determined, the required curve can be drawn.

Parabola by the Construction of a Diagram.—On the span RR', Fig. 27, describe an isosceles triangle whose height is double that of the bending moment. Divide the two sides AR and AR' of the triangle into any number of equal parts, and draw lines as in the figure.

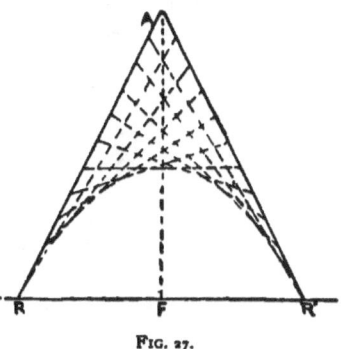

FIG. 27.

These lines will be tangents to the parabola, which may then be drawn.

Graphical Representation of Bending Moments and Shearing Forces for a Uniformly Distributed Load.—A uniform load may be considered as a system of equal and equidistant loads close together.

Thus in Fig. 28 the load area may be divided into any number of equal parts. The area of each part we may consider as the load which acts at its centre of gravity, and lay it off to any convenient scale in the force polygon, as at 1, 2, 3, 4, . . . 19, 20.

Since the reactions at R and R' are equal, we take the pole O in a horizontal through the centre of force line PP', and draw the radii $O1$, $O2$, $O3$, $O4$, . . . $O20$. Then describe the funicular polygon ab, bc, cd, de, . . . vw, by drawing ab parallel to PO, bc to $O1$, cd to $O2$, de to $O3$, . . . vw to $O20$.

The funicular polygon is now closed by the line aw, and a line OS is drawn through pole O (on OH in this example), which is parallel to aw and to the girder. Any ordinate taken to this parabola from the base line aw multiplied by the distance OH will give the bending moment at any point in the girder.

To avoid any error in direction carried on by the lines being too short, one half the number of divisions will be quite sufficient to get the moments.

The shearing forces on the webs are shown similar to the

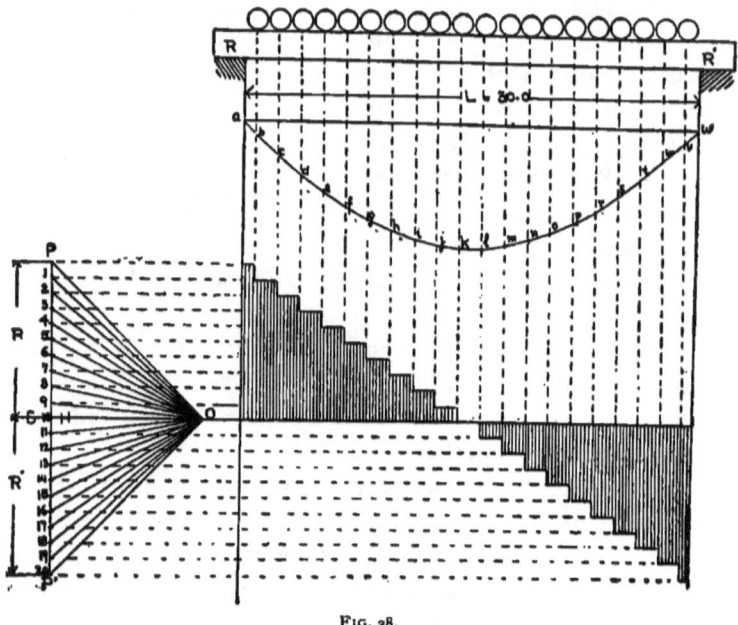

Fig. 28.

previous examples, and are equal to the distances of the points of the polygon of forces from S.

Therefore the shearing forces are taken from the force polygon used as ordinates as shown in the diagram, and the hatched figure is the result.

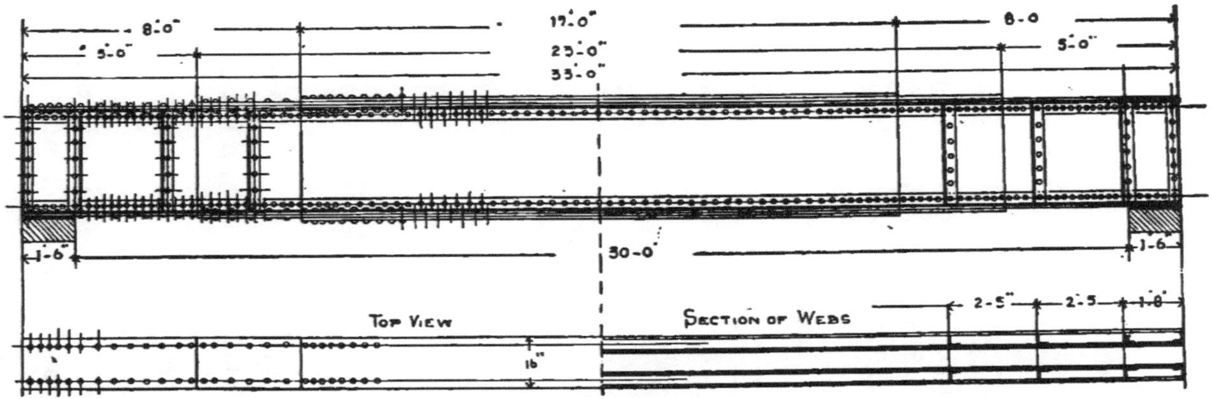

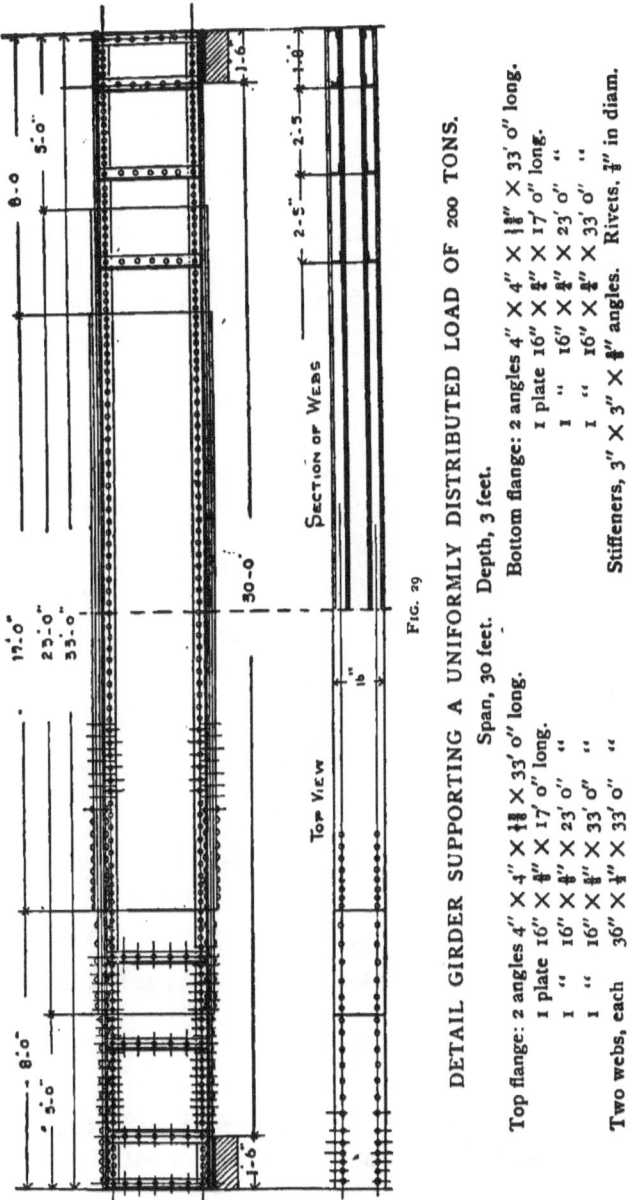

FIG. 29

DETAIL GIRDER SUPPORTING A UNIFORMLY DISTRIBUTED LOAD OF 200 TONS.

Span, 30 feet. Depth, 3 feet.

Top flange: 2 angles $4'' \times 4'' \times \frac{13}{16} \times 33'\ 0''$ long.
1 plate $16'' \times \frac{5}{8}'' \times 17'\ 0''$ long.
1 " $16'' \times \frac{5}{8}'' \times 23'\ 0''$ "
1 " $16'' \times \frac{5}{8}'' \times 33'\ 0''$ "
Two webs, each $36'' \times \frac{1}{2}'' \times 33'\ 0''$ "

Bottom flange: 2 angles $4'' \times 4'' \times \frac{13}{16}'' \times 33'\ 0''$ long.
1 plate $16'' \times \frac{3}{4}'' \times 17'\ 0''$ long.
1 " $16'' \times \frac{3}{4}'' \times 23'\ 0''$ "
1 " $16'' \times \frac{3}{4}'' \times 33'\ 0''$ "
Stiffeners, $3'' \times 3'' \times \frac{3}{8}''$ angles. Rivets, $\frac{7}{8}''$ in diam.

Example IV.

GIRDER SUPPORTING TWO CONCENTRATED LOADS.

In a girder supported at both ends, the bending moment at any point produced by *two* loads is the sum of the moments produced at that point by each load separately.

Draw the triangles Fig. 30, having vertices at *f* and *h*, repre-

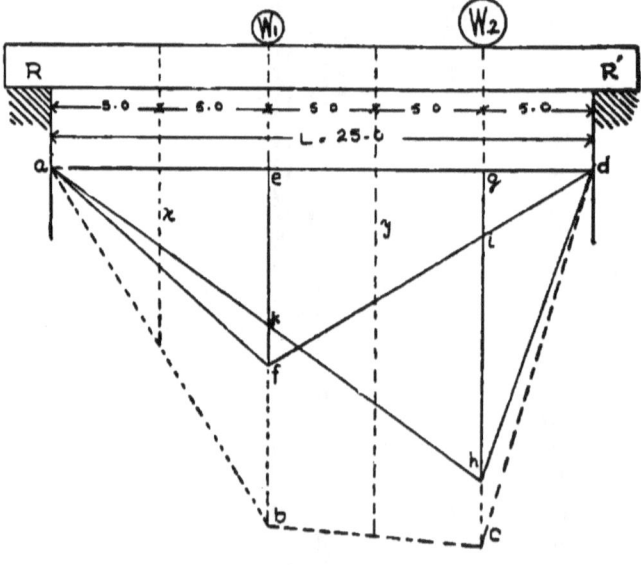

Fig. 30.

senting bending moments for loads W_1 and W_2, as in Fig. 20, by

$$M = W \times \frac{a \times b}{L}.$$

Extend *ef* to *b*, and *gh* to *c*, making each long vertical equal to the sum of each moment at the position of each load, or *eb*

EXAMPLE IV.   57

equal to $ef + ek$, and gc equal to $gh + gi$. Any ordinate, as x or y, measured from the base line ad to the polygonal figure $abcd$, will give by the same scale the moments at the corresponding points in the girder.

Or the bending moment for

$$W_1 \text{ at } W_2 = W_1 \times \frac{ae \times dg}{L};$$

$$W_2 \text{ at } W_1 = W_2 \times \frac{dg \times ae}{L}.$$

Example: What metal area would be required in the flanges of a *box girder* of 25 feet span 2 feet in depth to sustain 30 tons concentrated 10 feet from left support, and 70 tons 5 feet from right support?

Here $W_1 = 30$ tons, $W_2 = 70$ tons, $L = 25$ feet,

$d = 2$ feet, $A = $ flange area, $s = 6$ tons.

At e for W_1, $M = 30 \times \dfrac{10 \times 15}{25} = 180$ ton-feet.

At g for W_2, $M = 70 \times \dfrac{20 \times 5}{25} = 280$ ton-feet.

Draw the vertices ef and gh by scale equal to 180 and 280 ton-feet, respectively, and connect each to a and d; extend ef (equal to ek) to b, gh (equal to gi) to c, and connect a, b, c, d. Then eb and gc measured by the same scale will represent the bending moments produced at e and g by W_1 and W_2.

At x, $M = 160$ ton-feet,

$$A = \frac{160}{2 \times 6} = 13.33 \text{ square inches.}$$

At e, $M = 320$ ton-feet,

$$A = \frac{320}{2 \times 6} = 26.66 \text{ square inches.}$$

At y, $M = 330$ ton-feet,

$$A = \frac{330}{2 \times 6} = 27.5 \text{ square inches.}$$

At g, $M = 340$ ton-feet,

$$A = \frac{340}{2 \times 6} = 28.33 \text{ square inches.}$$

It will be noticed that the maximum bending moment by the diagram is under the load W_2. As a check upon the scale figures,

At W_2 for W_2, $M = W_2 \times \dfrac{a \times b}{L} = 70 \times \dfrac{5 \times 20}{25} = 280$ ton-feet.

At W_2 for W_1, $M = W_1 \times \dfrac{ae \times dg}{L} = 30 \times \dfrac{10 \times 5}{25} = 60$ "

Total at W_2 or at $g = \overline{340}$ "

Construction of Flanges.—The maximum bending moment being under the greatest load, the greatest amount of metal will be required in the flanges at that point. Then to make up the proper flange section:

Top flange = 2 angles $6'' \times 4'' \times \frac{7}{16}''$ = 8.36 square inches.
    1 plate $16'' \times \frac{1}{2}''$ = 8.00 " "
    2 plates $16'' \times \frac{3}{8}''$ = 12.00 " "

Total, $\overline{28.36}$ " "

EXAMPLE IV.   59

For the bottom flange, deduct rivet-holes as in our previous examples. Then using $\frac{7}{8}$-inch-diameter rivets, and allowing $\frac{1}{8}$ inch more: for one hole use $\frac{7}{16}'' + \frac{1}{2}'' + \frac{8}{8}'' + \frac{8}{8}'' \times 1'' = \frac{27}{16}$ inches, for two holes $\frac{27}{16} \times 2 = 3.38$ square inches, to be added to the bottom flange at g, or $28.33 + 3.38 = 31.71$ square inches.

Bottom flange = 2 angles $6'' \times 4'' \times \frac{7}{16}''$ = 8.36 square inches.
      1 plate $16'' \times \frac{9}{16}''$ = 9.00   "   "
    1   " $16'' \times \frac{1}{2}''$ = 8.00   "   "
    1   " $16'' \times \frac{3}{8}''$ = 6.00   "   "
         Total,   31.36   "   "

Flanges reduced in Area towards the Supports.—Construct the diagram Fig. 31 upon the span RR', making the

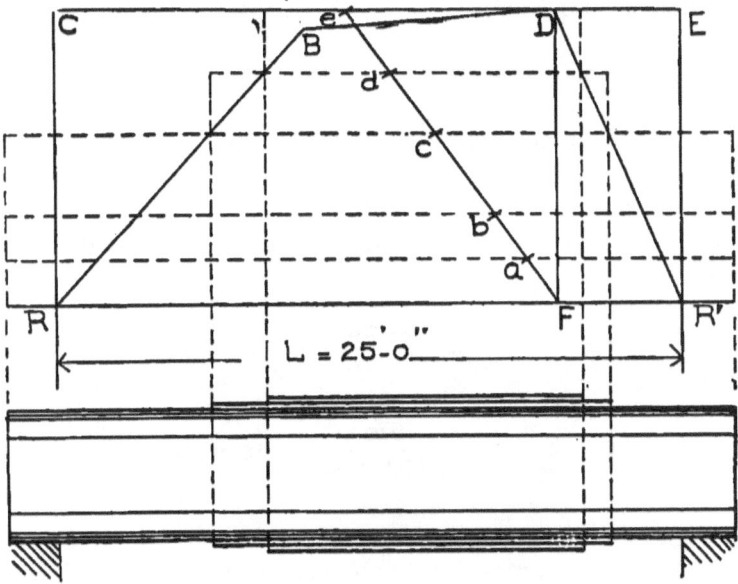

FIG. 31.

polygon $RBDR'$ similar to the bending-moment polygon, Fig. 30. At F, 5 feet from right support, make FD equal to the maximum bending moment, 340 ton-feet. Draw the rectangle $RCER'$. From F place the scale at any angle, as at Fe, until it measures 28.36 square inches.

For two angles set off 4.18 square inches each at a and b; one plate $16 \times \frac{1}{2} = 8$ square inches at c; two plates $16 \times \frac{3}{8} = 6$ square inches each at d and e. Horizontal lines drawn from $a, b, c, d,$ and e to the polygon $RBDR'$ and carried down to the base line RR' will give the position of the plates in each flange. Being a *box girder*, the angles and adjoining plates are required to extend the full length, all other plates 12 inches beyond the calculated area, to reach at least two cross-lines of rivets.

Webs.—To find the reaction at R', the right support, the centre of the moments is taken at the left support. In like manner to find the reaction at R, the left support, the centre of the moments is taken at the right support.

Then R' supports $(30 \times 10) + (70 \times 20) \div 25 = 68$ tons,

and R supports $(70 \times 5) + (30 \times 15) \div 25 = 32$ tons.

As a check, the sum of R and R' is seen to be 100 tons.

The thickness of web may then be determined by the formula:

At R', $T = \dfrac{e}{dk} = \dfrac{136000}{24 \times 6000} = .94$ or $\frac{15}{16}$ inch.

At R, $T = \dfrac{64000}{21 \times 6000} = .44$ or $\frac{7}{16}$ inch.

But as we have a box girder and two webs, adopting the greater thickness, each will be one half of $\frac{15}{16}$ or $\frac{15}{32}$, say $\frac{1}{2}$ inch, for the thickness; this thickness to extend the entire length, as explained under example of "One concentrated load not in centre."

EXAMPLE IV.

Stiffeners.—To determine whether we require stiffeners:
Safe resistance to buckling

$$= \frac{10000}{1 + \frac{24 \times 24}{3000 \times \frac{1}{2} \times \frac{1}{2}}} = 5656 \text{ lbs. per square inch.}$$

As this is almost what we have adopted for shearing (6000 pounds), the web will have to be stiffened at the bearings and under each concentrated load. A $3 \times 3 \times \frac{3}{8}$ inch angle on the outside of the web will be sufficient for the purpose (see girder drawing, Fig. 33).

Rivets.—The shearing area being less than the bearing, the former will have to be adopted.

The area of a $\frac{7}{8}$-inch-diameter rivet $= \frac{7}{8} \times \frac{7}{8} \times .7854 = .6013$. This multiplied by $7500 = 4510$ pounds, safe shear for each rivet. The maximum bending moment at g.

$M = 340$ ton-feet, and divided by the depth, the horizontal strain $= \frac{340}{2} = 170$ tons or 340,000 pounds; and then divided by $4510 = 75$ rivets, to be placed a distance of 60 inches for one web or 120 inches for both, spaced about $1\frac{5}{8}$ inches centres from position of maximum M to R' support. As this is less than the minimum spacing in a straight line previously adopted, the rivets will require to be staggered. This is also the practical reason why large angles, with the longer leg vertical, became necessary.

By referring to the graphical shearing-force diagram, Fig. 32, it will be seen that the horizontal increments of strain in the web are uniform from g to R'; therefore rivets will be spaced $1\frac{5}{8}$ centres that distance.

In dividing the 75 rivets in the distance from g to e it will be noticed in the diagram that there is little shear in the webs between e and g. We will therefore space the rivets the maximum adopted, or 6 inches centres.

The rivet spacing between e and R will have to be spaced by the horizontal strain at that point.

At e, $M = 320$ ton-feet, and divided by the depth, the horizontal strain $= \frac{320}{2} = 160$ tons or 320,000 pounds; and then divided by 4510 $= 70$ rivets, to be placed a distance of 120 inches for one web, 240 inches for both, spaced about $3\frac{7}{10}$ inches from e to R support.

Graphical Representation of the Bending Moments and Shearing Forces in a Girder supporting Two Concentrated Loads.—We have in the example a girder of 25 feet span sustaining a concentrated load of 30 tons 10 feet from left support

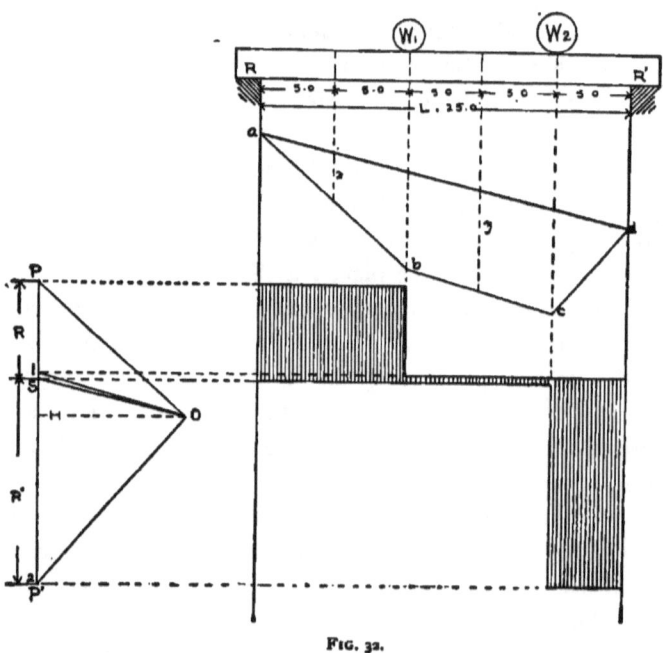

FIG. 32.

and 70 tons 5 feet from right support. We shall first determine R and R', the pressure on the supports, and then the vertical or transverse stresses.

EXAMPLE IV.

Set the given forces W_1 and W_2 off in succession along the line PP', Fig. 32, W_1 of 30 tons at 1, W_2 of 70 tons at 2. The line PP' is thus the polygon of the given forces W_1, W_2, and $P'P$, its closing line, is their resultant.

Take any point O as pole, equal to ten units of the scale adopted, and draw the radii OP, $O1$, $O2$. Then describe the funicular polygon $abcd$ by drawing ab parallel to OP, terminating in W_1 produced; bc parallel to $O1$, terminating in W_2; and cd parallel to O_2, terminating in the prolongation downwards of R'. The funicular polygon is now closed by the line da, and a line OS is drawn through the pole O parallel to da.

Then as a condition of equilibrium,

the reactions at $R = PS$, and at $R' = SP'$.

Any ordinate from ad in the funicular polygon, as xy, measured to the inclined line ab, bc, or cd, multiplied by the pole distance OH, will give the bending moments at any point in the girder.

The shearing forces are equal to the distances of the various points of the polygon of forces from S.

Accordingly the hatched figure gives the shearing forces at any section of the girder; and the shear on the webs can be measured by the same scale.

The shear on the web from R' support to the 70-ton load W_2 is uniform, as is also the 30-ton load W_1 to R support. Between the two loads it will be noticed that there is but little shearing force.

COMPOUND RIVETED GIRDERS.

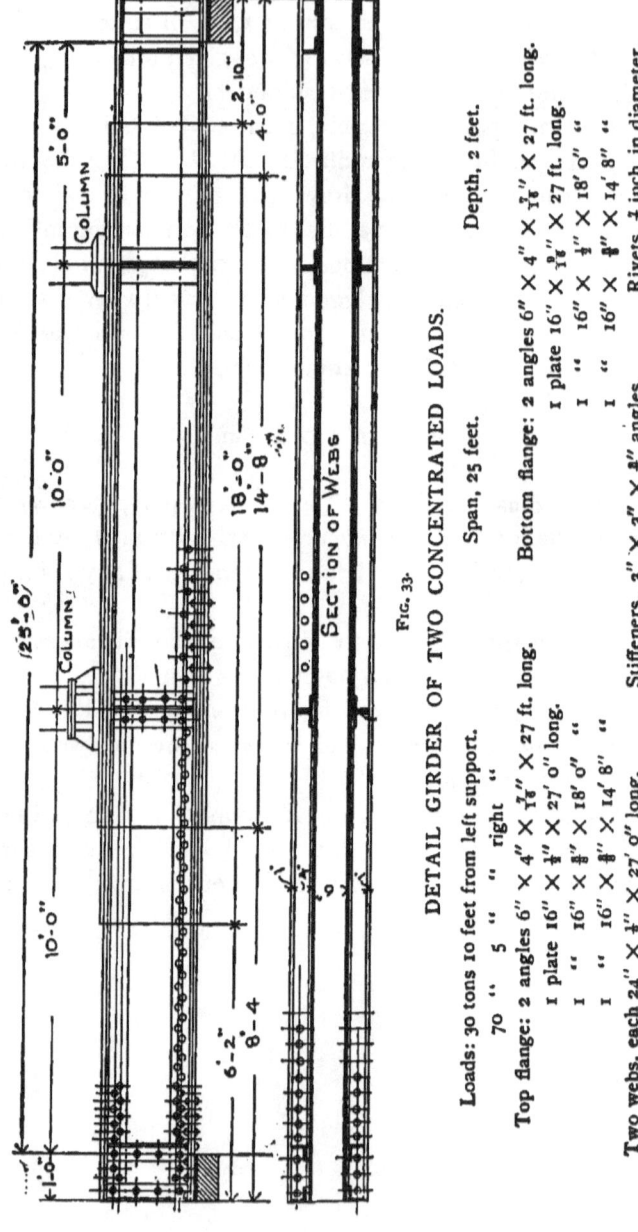

Fig. 33.

DETAIL GIRDER OF TWO CONCENTRATED LOADS.

Loads: 30 tons 10 feet from left support.   Span, 25 feet.   Depth, 2 feet.
      70 " 5 " " right "

Top flange: 2 angles $6'' \times 4'' \times \tfrac{7}{16}'' \times 27$ ft. long.   Bottom flange: 2 angles $6'' \times 4'' \times \tfrac{7}{16}'' \times 27$ ft. long.
      1 plate $16'' \times \tfrac{1}{2}'' \times 27'$ 0" long.   1 plate $16'' \times \tfrac{9}{16}'' \times 27$ ft. long.
      1 " $16'' \times \tfrac{1}{2}'' \times 18'$ 0" "   1 " $16'' \times \tfrac{1}{2}'' \times 18'$ 0" "
      1 " $16'' \times \tfrac{3}{8}'' \times 14'$ 8" "   1 " $16'' \times \tfrac{3}{8}'' \times 14'$ 8" "

Two webs, each $24'' \times \tfrac{1}{4}'' \times 27'$ 0" long.   Stiffeners, $3'' \times 3'' \times \tfrac{3}{8}''$ angles.   Rivets, $\tfrac{3}{4}$ inch in diameter.

Example V.

GIRDER SUPPORTING TWO CONCENTRATED LOADS AND A UNIFORMLY DISTRIBUTED LOAD.

In a girder supported at both ends, the bending moments at any point produced by all the loads is the sum of the moments produced at that point by each of the loads separately. This is a combination of the two previous examples, and each is to be taken separately.

The polygon for the concentrated loads to be drawn under, and the parabola for the uniform load over, the girder.

Example: What metal area would be required in the flanges of a box girder 16 inches wide, 32 feet span, 3 feet in depth, to sustain 60 tons concentrated 10 feet from right support, 40 tons 10 feet from left support, and a uniformly distributed load of 80 tons, with 6 tons unit strain per square inch in the flanges?

Draw the triangles Fig. 34, having vertices at f and h, representing bending moments by

$$M = W \times \frac{a \times b}{L} \text{ for loads } W_1 \text{ and } W_2 \text{ (Example II).}$$

At e for W_1, $M = 40 \times \dfrac{10 \times 22}{32} = 125$ ton-feet.

At g for W_2, $M = 60 \times \dfrac{10 \times 22}{32} = 412.5$ ton-feet.

Draw the vertices ef and gh by scale equal to 125 and 412.5 ton-feet respectively, and connect each to ad; extend ef (equal to ek) to b, and gh (equal to gi) to c, and connect a, b, c, d.

Then *eb* and *ge* measured by same scale will give the bending moments produced at *e* and *g* by W_1 and W_2.

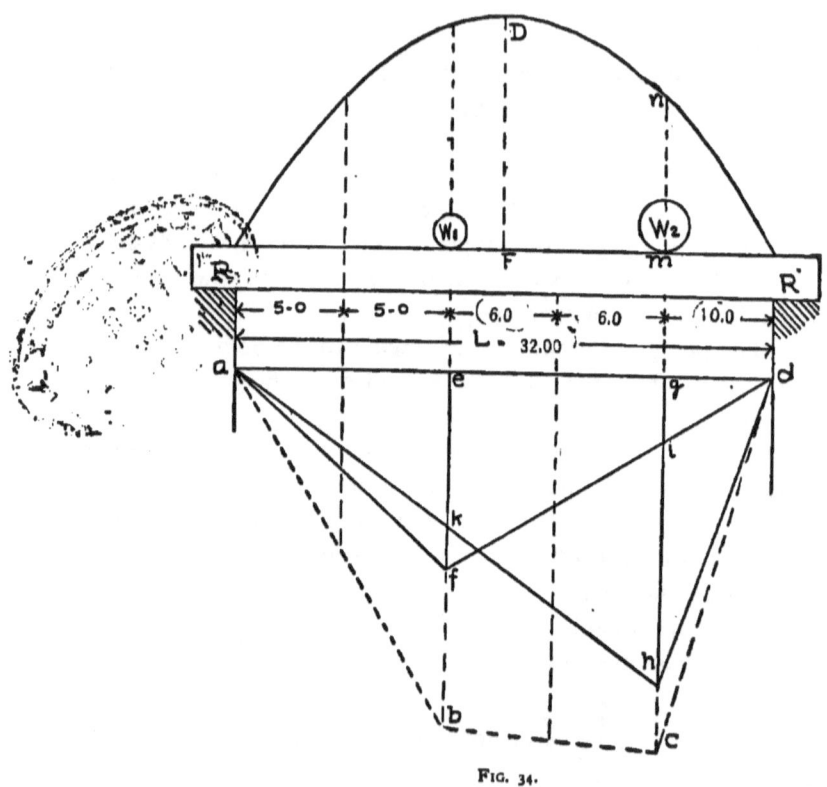

Fig. 34.

Draw the parabola for the uniform load, making *FD*, by the same scale as for the concentrated loads, equal to the moment at the centre of the girder by formula:

$$\text{At centre, } M = \frac{WL}{8} = \frac{80 \times 32}{8} = 320 \text{ ton-feet.}$$

The bending moment due to the two concentrated loads and the uniform load at any point in the girder is equal at that

EXAMPLE V.

point to the sum of the ordinates of the parabola RDR' and the polygon $abcd$.

Then at W_2, the ordinate of the parabola mn and the polygon gc,

$$M = 812.5 \text{ ton-feet.}$$

Or at g, for W_2, $M = 60 \times \dfrac{22 \times 10}{32} = 412.5$ ton-feet;

* " W_1, $M = 40 \times \dfrac{10 \times 10}{32} = 125.0$

† for uniform load $= \dfrac{\frac{80}{16} \times 10}{2} \times 22 = 275.00$

Maximum $M = 812.5$ "

Flange area $= \dfrac{812.5}{3 \times 6} = 45.12$ square inches.

Construction of Flanges.—Then to make up the 45.12 square inches in the flanges at g, we would require:

Top flange = 2 angles $6'' \times 4'' \times \frac{5}{8}'' = 11.72$ square inches.
  1 plate $16'' \times \frac{3}{4}'' = 12$ "
  2 plates $16'' \times \frac{11}{16}'' = 22.00$ "

  Total, 45.72 "

For the bottom flange we deduct rivet-holes, then using ⅞-inch-diameter rivets and allowing ⅛ inch more, we have for one hole $\frac{10}{16} + \frac{12}{16} + \frac{11}{16} + \frac{11}{16} \times 1 = 2\frac{12}{16}$ inches, for two holes $2\frac{12}{16} \times 2 = 5\frac{1}{2}$ square inches, to be added to the bottom flange at g, or $45.12 + 5.525 = 50.62$ square inches.

---

* Refer to formula under girder of one concentrated load not at centre.
† Refer to formula under girder of a uniformly distributed load.

COMPOUND RIVETED GIRDERS.

Bottom flange = 2 angles $6'' \times 4'' \times \frac{5}{8}''$ = 11.72 sq. in.
1 plate $16'' \times \frac{7}{8}''$ = 14.00 " "
1 " $16'' \times 1\frac{3}{16}''$ = 13.00 " "
1 " $16'' \times \frac{3}{4}''$ = 12.00 " "

Total, 50.72 " "

Webs.—The reactions due to both uniform and concentrated loads may be obtained by adding together the reactions due to the uniform load and each concentrated load, or they may be computed in one operation.

To find the right reaction, R', the centre of moments is taken at the left support and the uniform load regarded as concentrated at its middle; then the equation of moments is

$$R' \times 32 = 60 \times 22 + 40 \times 10 + 80 \times 16,$$

from which $R' = 93.75$ tons. In like manner, to find R the centre of moments is taken at the right support and

$$R \times 32 = 60 \times 10 + 40 \times 22 + 80 \times 16,$$

from which $R = 86.25$. As a check the sum of R' and R is seen to be 180 tons, which is the same as the sum of the two concentrated loads and the uniform load.

The thickness of web may then be determined by the formulas:

At R', $t = \dfrac{G}{dk} = \dfrac{187500}{36 \times 6000} = .868 = \frac{7}{8}$ inch.

At R, $t = \dfrac{172560}{36 \times 6000} = .798 = 1\frac{3}{8}$ "

Adopting the greater thickness and having two webs, each will be one half of $\frac{7}{8}$ or $\frac{7}{16}$.

EXAMPLE V.

Stiffeners.—To determine whether we require stiffeners:
Safe resistance to buckling

$$= \frac{10000}{1 + \frac{36 \times 36}{3000 \times \frac{7}{16} \times \frac{7}{16}}} = 3069 \text{ lbs. per square inch.}$$

But having adopted 6000 lbs. per square inch as the safe shear, we will require stiffeners to be placed throughout the length, from W_4 to R and W_{12} to R', spaced about 3 ft. centres. No stiffeners are required between W_4 and W_{12}, there being little shear on the webs in that length. Refer to the diagram Fig. 35.

Then to stiffen the web against buckling, place a 4″ × 4″ × ½″ angle outside of web. We then have for the thickness $\frac{7}{16}$ + ½ = $\frac{15}{16}$ inch, and the formula becomes:

Safe resistance to buckling

$$= \frac{10000}{1 + \frac{36 \times 36}{3000 \times \frac{15}{16} \times \frac{15}{16}}} = 6700 \text{ lbs. per square inch.}$$

Rivets.—The shearing area is again used in this example and ⅞-inch-diameter rivets.

The safe single shear for each ⅞ rivet by table = 4510 pounds per square inch. The maximum bending moment is at g, and $M = 812.5$ ton-feet. Then divided by the depth, the horizontal strain $= \frac{812.5}{3} = 270.2$ tons or 540,400 pounds. This divided by 4510 = 119 rivets, to be placed a distance of 120 inches for one web or 240 inches for both, spaced about 2 inches centres (staggered) from position of maximum M to R' support, from g to e. It will be noticed in the diagram Fig. 35 that there is but little sheer on the web; therefore the maximum (6 inches) spacing should be adopted.

The rivet spacing from e to R will be regulated by the horizontal strain at that point.

At e, $M = 737.625$ ton-feet.

Horizontal strain $= \dfrac{737.625}{3} = 245.875$ tons or $491,750$ pounds,

and then divided by $4510 = 109$ rivets, to be placed a distance of 120 inches for one or 240 inches for both webs, spaced about $2\frac{1}{4}$ inches (staggered) from e to R support.

Graphical Representation for Two Concentrated Loads and a Uniform Load.—In the following diagram, Fig. 35, we

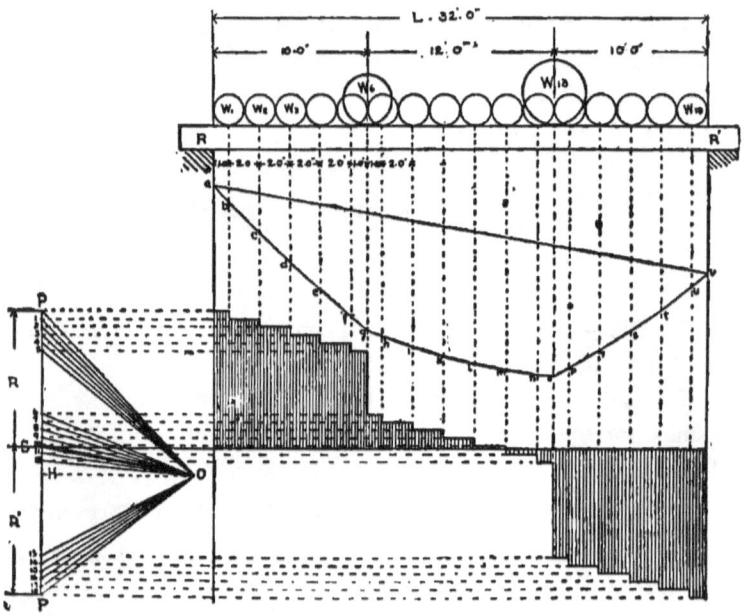

Fig. 35.

have a combination of the previous examples. The uniform load to be considered is a system of equal and equidistant loads

EXAMPLE V.

close together, as in Fig. 28. In determining the reaction on the support, we set the given loads W_1, W_2, W_3, ... W_{16} off in succession along the line PP'. The uniform load of 80 tons being divided into 16 parts, each equal to 5 tons, set W_1 of 5 tons at 1; W_2 of 5 tons at 2; W_3 of 5 tons at 3; W_4 of 5 tons at 4; W_5 of 5 tons at 5; W_6, the concentrated load, at 6; and so on to the end. The line PP' is then the polygon of the forces, and $P'P$, the closing line, is the resultant.

Take any point O as a pole equal to ten units of the scale adopted, and draw the radii OP, $O1$, $O2$, $O3$, $O4$, $O5$, ... $O18$. Then describe the funicular polygon $abcd$ ... $stuv$ by drawing ab parallel to OB, bc parallel to $O1$, terminating in the prolongation of W_1; cd parallel to $O2$, terminating in W_2 produced; ... finally, uv parallel to $O18$, terminating in the prolongation downwards of R'.

The funicular polygon is now closed by the line va, and a line OS is drawn through the pole O parallel to va.

Then as a condition of equilibrium

the reaction at $R = PS$, and at $R' = SP'$.

Any ordinate from av in the funicular polygon and measured to the inclined lines ab, bc, cd, de, ... uv, multiplied by the pole distance OH, will give the bending moment at any point in the girder.

The shearing forces on the web can be measured on the hatched figure as explained in our previous examples. The greatest shear is at the bearings, and extends from the concentrated loads W_6 and W_{11} to R and R'. There is little or no shear on the web in centre of the girder.

The Area of Flanges reduced by the Funicular Polygon.—Construct the diagram Fig. 36, by drawing the polygon RDR' similar to the funicular polygon Fig. 35.

From a point on the closing line RR' directly under the

maximum bending moment at *F*, draw *FD* at the same point; place the scale at any angle until it meets *DD'* perpendicular to *FD* at *e*, measuring 45.12 square inches of the top flange.

For the two angles set off 5.86 square inches each at *a* and *b*, then one plate 16 × ¾ or 12 square inches at *c*, two plates

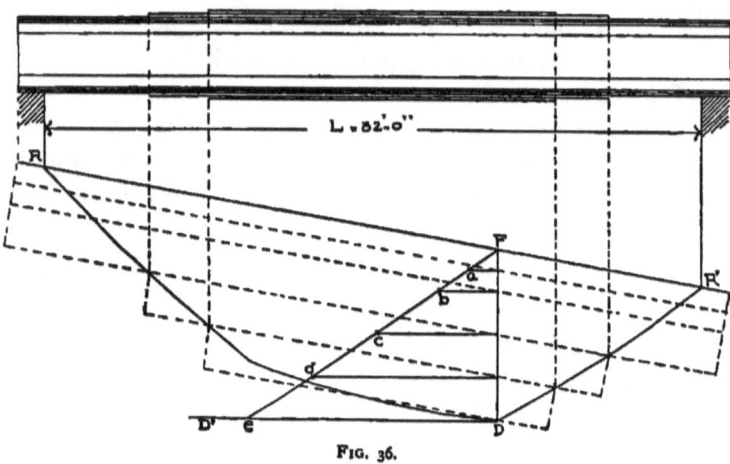

FIG. 36.

16 × 11/16 or 11 square inches each at *d* and *e*. Horizontal lines drawn to *FD*, and again lines drawn from *FD* parallel to *RR'*, intersecting the polygon and carried up to *RR'*, will give the position of the plate in each flange. The angles and adjoining plates to extend the full length of girder. The other plates to extend over the calculated distance to reach at least four cross-lines of rivets.

EXAMPLE V.

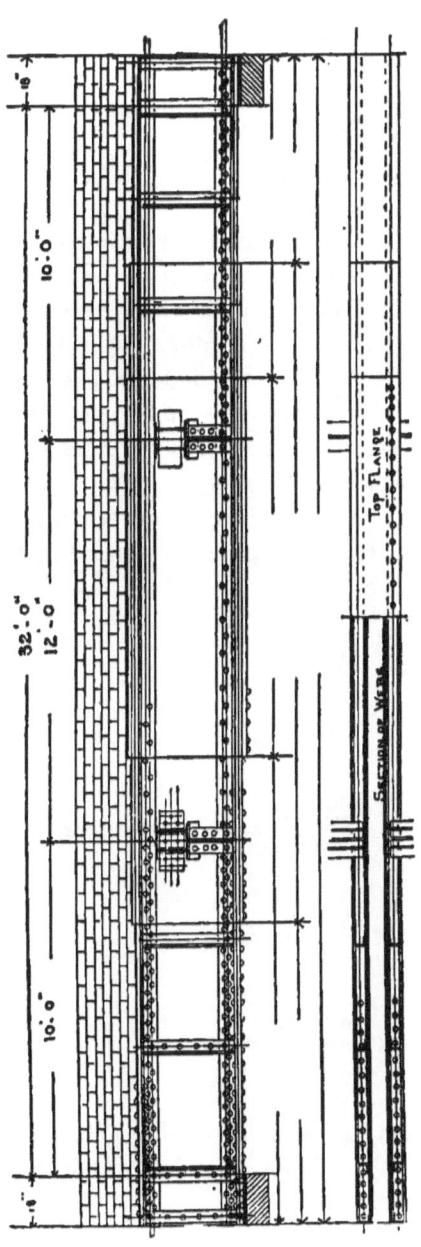

FIG. 37.

DETAIL GIRDER SUPPORTING TWO CONCENTRATED LOADS AND A UNIFORMLY DISTRIBUTED LOAD.

Loads: 60 tons concentrated 10' 0" from left support. Span, 32 feet. Depth, 3 feet.
       40 "   "   "   "   right "
       80 "   uniformly distributed.

Top flange: 2 angles 6" × 4" × ⅜". Bottom flange: 2 angles 6" × 4" × ⅜".
            1 plate 16" × ⅜".   1 plate 16" × ⅜".
            1 " 16" × 1⅜".   1 " 16" × 1⅜".
            1 " 16" × 1⅛".   1 " 16" × ⅜".

Webs, 7/16" × 3' 36". Stiffeners, 4" × 4' × ½" angles. Rivets, ⅞ inch diameter.

Example VI.

Girder Supporting Three Concentrated Loads.

In a girder supported at both ends, the bending moment at any point produced by three concentrated loads is the sum of the moments produced at that point by each load separately.

Example: What metal area would be required in the flanges of a *box girder* of 35 feet span, 2 feet 6 inches in depth, to sustain 40 tons concentrated 5 feet and 70 tons 15 feet from left support, and 60 tons concentrated 10 feet from right support, the girder to be 20 inches wide?

Draw the triangles Fig. 38 as in the previous examples, having vertices at t, i, and k, representing bending moments by

$$M = W \times \frac{a \times b}{L}$$

for loads W_1, W_2, and W_3, as in Example II.

At H, for W_1, $M = 40 \times \dfrac{5 \times 30}{35} = 171.43$ ton-feet.

At f, for W_2, $M = 70 \times \dfrac{15 \times 20}{35} = 600.00$ "

At g, for W_3, $M = 60 \times \dfrac{10 \times 25}{35} = 428.57$ "

Draw the vertices ht, fi, and gk by scale equal to 171.43, 600, and 428.57 ton-feet respectively. Connect each to R and R'.

EXAMPLE VI.

Extend *ht* to *b*, equal to the sum of *hr* and *hs*; extend *fi* to *c*, equal to the sum of *fo* and *fn*; extend *gk* to *d*, equal to the sum of *gl* and *gm*. Connect *a*, *b*, *c*, and *d*. Then *hb*, *fc*, and *gd*,

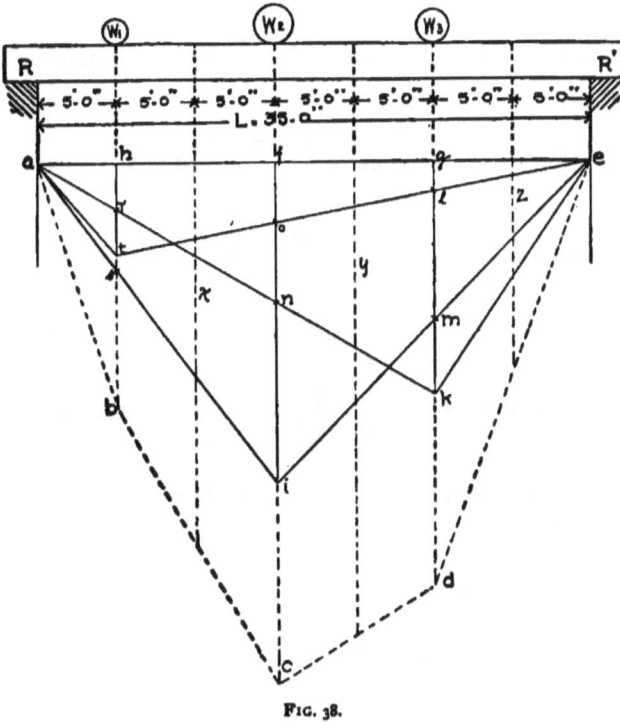

FIG. 38.

measured by same scale, will represent the bending moments produced at *h*, *f*, and *g* in the girder by loads W_1, W_2, and W_3.

At any point in the girder, as *x*, *y* or *z* will, by same scale measured from base line *ae* to the polygon *abcde*, represent the bending moments at their respective points.

Then by scale:

At *h*, $M = 457.13$ ton-feet.   At *y*, $M = 878.57$ ton-feet.
" *x*, " $= 714.28$ "   " g_1, " $= 785.71$ "
" *f*, " $= 971.42$ "   " *z*, " $= 392.85$ "

From the above diagram it will be observed that the *maximum* M occurs at point f. For the proof refer to Fig. 20, diagram of one concentrated load not at centre.

$$\text{At } f, \text{ for } W_3, M = 70 \times \frac{15 \times 20}{35} = 600 \quad \text{ton-feet.}$$

$$\text{At } f, \text{ for } W_1, M = 40 \times \frac{5 \times 20}{35} = 114.28 \quad \text{"}$$

$$\text{At } f, \text{ for } W_2, M = 60 \times \frac{10 \times 15}{35} = 257.14 \quad \text{"}$$

$$\text{Max. } M = 971.42 \quad \text{"}$$

$$\text{Area of flange} = \frac{971.42}{2.5 \times 6} = 64.76 \text{ square inches.}$$

Construction of Flanges.—To make up the 64.76 square inches in the flanges at f we would require:

Top flange = 2 angles $6'' \times 4'' \times \frac{5}{8}'' = 11.72$ square inches.
    1 plate $20'' \times \frac{3}{4}''$ = 15.0 " "
    3 plates $20'' \times \frac{5}{8}''$ = 37.5 " "

        Total, 64.22 " "

.54 of an inch less than required.

For the bottom flange we deduct rivet-holes. Then using $\frac{7}{8}$-inch-diameter rivets and allowing $\frac{1}{8}$ inch more, we have for one hole $\frac{5}{8}'' + \frac{3}{4}'' + \frac{5}{8}'' + \frac{5}{8}'' + \frac{5}{8}'' \times 1'' = \frac{26}{8}$, for two holes $\frac{26}{8} \times 2 = \frac{52}{8} = 6\frac{4}{8}$ square inches. On account of the closeness of the rivets it will be noticed, by referring to the girder drawing Fig. 41, another hole in the vertical leg of the angle will require to be deducted; we have then for one hole $\frac{5}{8} \times 1 = \frac{5}{8}$, for two holes $\frac{5}{8} \times 2 = \frac{10}{8} = 1\frac{1}{4}$ square inches, to be added

EXAMPLE VI.

to the bottom flange at f, or $64.76 + 6.375 + 1.25 = 72.385$ square inches. Then

Bottom flange: 2 angles $6'' \times 4'' \times \frac{5}{8}'' = 11.72$ sq. inches.
4 plates $20'' \times \frac{3}{4}'' = 60$ " "

Total, $ 71.72$ " "

Webs.—To find the right reaction at R', the centre of moments is taken at the left support. Then the equation of moments is

$$R' \times 35 = 40 \times 5 + 70 \times 15 + 60 \times 25,$$

from which $R' = \dfrac{2750}{35} = 78.57$ tons. In like manner to find R, the centre of moments is taken at the right support, and

$$R \times 35 = 60 \times 10 + 70 \times 20 + 40 \times 30,$$

from which $R = \dfrac{3200}{35} = 91.43$ tons. As a check, the sum of R' and R is seen to be 170 tons.

Then the thickness of the web becomes·

At R', $t = \dfrac{157140}{30 \times 6000} = .873$, nearly $\frac{7}{8}$ of an inch.

At R, $t = \dfrac{182860}{30 \times 6000} = 1.15$, say 1 inch.

Adopting the greater thickness and having two webs, each will be $\frac{1}{2}$ inch thick.

COMPOUND RIVETED GIRDERS.

Stiffeners.—To determine whether we require stiffeners:
Safe resistance to buckling

$$= \frac{10000}{1 + \frac{30 \times 30}{3000 \times \frac{3}{8} \times \frac{3}{8}}} = 4545 \text{ pounds per square inch.}$$

This being less than 6000 pounds, the safe shear per square inch, the webs will require stiffeners at bearings, under each concentrated load, and the intervening distances from R to W_1, W_1 to W_2, W_3 to R', every three feet. No stiffeners are required between W_2 and W_3. Refer to Diagram Fig. 40.

Then to stiffen web at the above-stated points, place a $4'' \times 4'' \times \frac{3}{8}''$ angle on outside of each web, and the formula becomes:

Safe resistance to buckling

$$= \frac{10000}{1 + \frac{30 \times 30}{3000 \times \frac{7}{8} \times \frac{7}{8}}} = 7184 \text{ pounds per square inch.}$$

Rivets.—The safe shearing area of a $\frac{7}{8}$-inch-diameter rivet from table = 4510 pounds.

The maximum bending moment at f:

$$M = 971.42 \text{ ton-feet.}$$

Horizontal strain $= \dfrac{971.42}{2.5} = 388.66$ tons or $777{,}320$ pounds.

This divided by 4510 = 170 rivets, to be placed a distance of 180 or 360 inches in both webs, spaced about $2\frac{1}{8}$ inches (staggered) from f to R support.

EXAMPLE VI.

The maximum bending moment at g:

$$M = 785.71 \text{ ton-feet.}$$

Horizontal strain $= \dfrac{785.71}{2.5} = 314.4$ tons or 629,000 pounds.

Then divided by 4510 = 139 rivets, to be placed a distance of 120 inches for one or 240 inches for both webs, spaced about $1\frac{3}{4}$ inches (staggered) from g to R'.

The rivets to be spaced between W_1 and W_2, 6 inches centres.

Flange Plates reduced in Area towards the Supports.—Draw the diagram Fig. 39, as in our previous examples, upon the span $R\,R'$, making the polygon $RBDGR'$ similar to the bending moment polygon, Fig. 38. At F, 15 feet from R support, draw FD equal to the maximum.

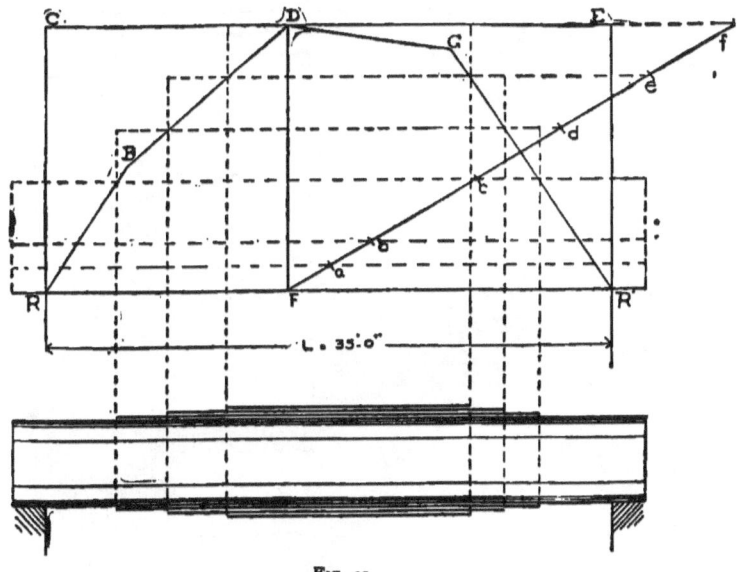

FIG. 39.

Bending moment due to the sum of the load W_1, W_2, and W_3 at that point, or 971.42 ton-feet. Draw the rectangle $RCER'$;

then from F place the scale at any angle, as at Ff, until it measures 64.76 square inches of the top flange. For two angles set off 5.86 square inches each at a and b; one plate 20″ × ¾″, or 15 square inches, at c; 3 plates 20″ × ⅝″, or 1.25 square inches, each at e, d, and f. Horizontal lines drawn from a, b, c, d, e, and f to the polygon and carried down to base line RR' will give the position of plates in each flange.

The angles and adjoining plates to extend the full length of girder. The plates to extend over the calculated lengths, equal to two cross-lines of rivets.

Graphical Representation of the Bending Moments and Shearing Forces for Three Concentrated Loads.— We have in this example a girder of 35 feet span, to sustain a

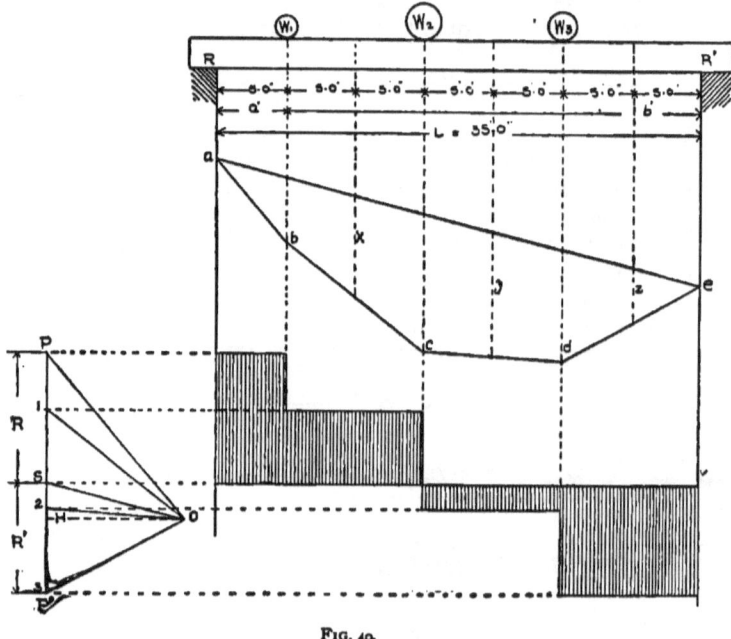

FIG. 40.

concentrated load of 40 tons 5 feet from R support, 70 tons 20 feet, and 60 tons 10 feet, from R' support. We shall first

EXAMPLE VI.

determine, as in our previous examples, the pressure on the supports.

Set the given loads W_1, W_2, and W_3, Fig. 40, off in succession along a line PP': W_1 of 40 tons at 1, W_2 of 70 tons at 2, and W_3 of 60 tons at 3. The line PP' is thus the polygon of the forces W_1, W_2, and W_3, and $P'P$, its closing line, is their resultant. Take any point O as pole, equal to ten units of the scale adopted, and draw the radii OP, $O1$, $O2$, $O3$. Then describe the funicular polygon $abcde$ by drawing ab parallel to OP, terminating in W_1 produced, bc parallel to $O1$, cd parallel to $O2$, terminating in the prolongation of W_2 and W_3 respectively, and de parallel to $O3$, terminating in the prolongation downwards of R'. The funicular polygon is now closed by the line ea, and a line OS is drawn through the pole O parallel to ea. Then as a condition of equilibrium,

the reaction at $R = PS$, and at $R' = SP'$.

The bending moments at any point in the girder can be measured by the ordinates of the funicular polygon multiplied by the pole distance OH. The shearing forces at any point on the webs to be measured from the hatched figures. The maximum shear on the webs is at R support, and is equal by scale to 91.43 tons, the same as previously calculated.

82 COMPOUND RIVETED GIRDERS.

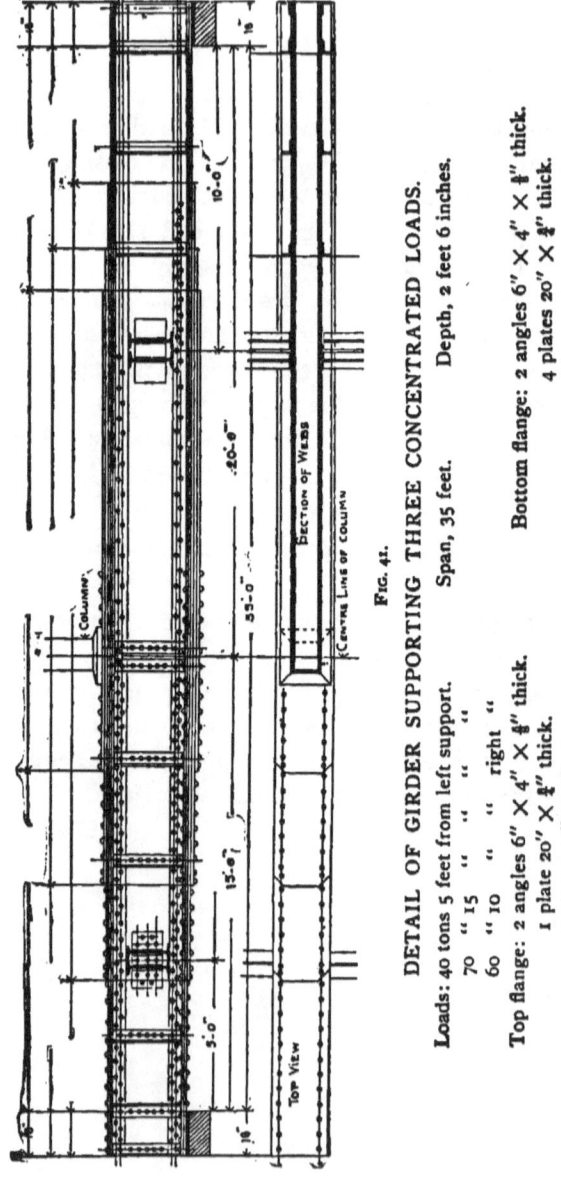

Fig. 41.

DETAIL OF GIRDER SUPPORTING THREE CONCENTRATED LOADS.

Loads: 40 tons 5 feet from left support.   Span, 35 feet.   Depth, 2 feet 6 inches.
70 " 15 " " " "
60 " 10 " " right "

Top flange: 2 angles 6" × 4" × $\frac{3}{8}$" thick.   Bottom flange: 2 angles 6" × 4" × $\frac{3}{8}$" thick.
1 plate 20" × $\frac{1}{2}$" thick.   4 plates 20" × $\frac{1}{2}$" thick.
3 plates 20" × $\frac{3}{8}$" "

Two webs, 30" × $\frac{1}{4}$" thick.   Stiffeners, 4" × 4" × $\frac{3}{8}$" angles.   Rivets, $\frac{7}{8}$ inch in diameter.

Example VII.

Girder Supporting Four Concentrated Loads.

The bending moment at any point produced by all the loads is the sum of the moments produced at that point by each of the loads separately, being the same as two and three concentrated loads.

Example: What metal area would be required in the flanges of a box girder of 40 feet span, 3 feet in depth, to sustain 20 tons concentrated 5 feet and 60 tons 15 feet from left support, 50 tons 10 feet and 30 tons 5 feet from right support, the girder to be 20 inches in breadth.

Draw the triangles, Fig. 42, having vertices at n, t, x, and q, representing bending moments by

$$M = W \times \frac{a \times b}{L}$$ for loads W_1, W_2, W_3, and W_4 (as in Ex. II).

At g, for W_1, $M = 20 \times \frac{5 \times 35}{40} = 87.5$ ton-feet.

At h, for W_2, $M = 60 \times \frac{15 \times 25}{40} = 562.5$ "

At i, for W_3, $M = 50 \times \frac{10 \times 30}{40} = 375$ "

At k, for W_4, $M = 30 \times \frac{5 \times 35}{40} = 131.25$ "

Draw the vertices gn, ht, ix, and kq by scale equal to 87.5, 562.5, 375, and 131.25 ton-feet respectively, and connect each with R and R'. Extend gn to b, equal to $gl + gm + go$; extend ht to c, equal to $hp + hr + hs$; extend ix to d, equal to $iu + iv + iw$; extend kq to e, equal to $kj + ky + kz$. Con-

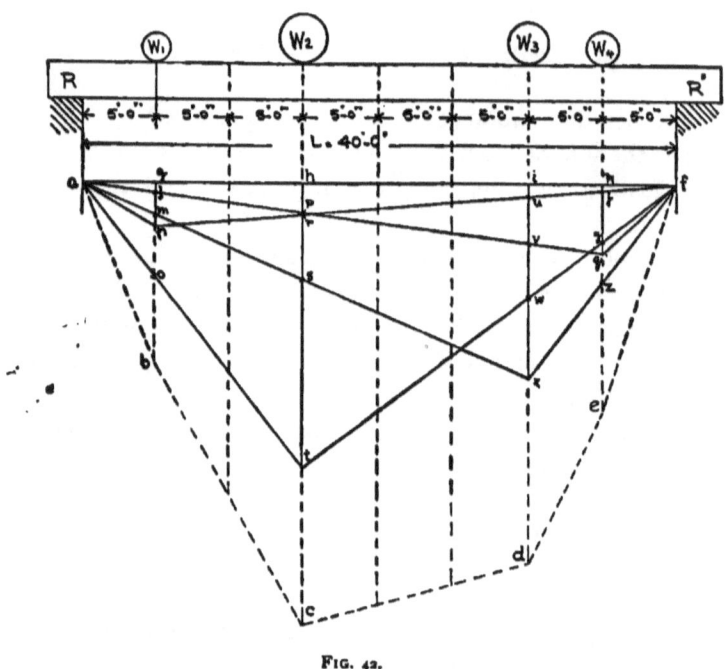

FIG. 42.

nect a, b, c, d, e, f. Then gb, hc, id, and ke, measured by same scale, will represent the bending moments produced at $g, h, i,$ and k in girder by loads $W_1, W_2, W_3,$ and W_4.

Then at g, $M = 356.25$ ton-feet; at i, $M = 781.25$ ton-feet. at h, $M = 868.75$ " at k, $M = 737.50$ "

The maximum bending moment due to the several loads it can be seen at a glance on the diagram is under the load at h,

EXAMPLE VII.

the point in the girder requiring the greatest amount of metal in the flanges.

At h, $M = 868.75$.

$$A = \frac{868.75}{3 \times 6} = 48.27 \text{ square inches in the top flange.}$$

Construction of Flanges.—To make up the required section we will construct the section as follows:

Top flange = 2 angles $5'' \times 3\frac{1}{2}'' \times \frac{11}{16}''$ = 10.74 sq. inches.
3 plates 20″ × ⅝″ = 37.5 "
Total, 48.24 "

For the bottom flange we deduct the rivet-holes, then using $\frac{7}{8}''$-diameter rivets and allowing $\frac{1}{8}''$ more, we have for one hole $\frac{11}{16}'' + \frac{5}{8}'' + \frac{5}{8}'' + \frac{5}{8}'' \times 1'' = \frac{41}{16}''$, for two holes $\frac{41}{16} \times 2 = \frac{82}{16} = 5\frac{1}{8}$ square inches, to be added to the bottom flange at h, or $48.24 + 5.125 = 53.365$ square inches.

Bottom flange = 2 angles $5'' \times 3\frac{1}{2}'' \times \frac{11}{16}''$ = 10.74 sq. inches.
2 plates 20″ × ¾″ = 30.00 "
1 " 20″ × ⅝″ = 12.5 "
53.24 "

Webs.—To find the shear on the web at R', the centre of moments is taken at R.

Then $R' \times 40 = 20 \times 5 + 60 \times 15 + 50 \times 30 + 30 \times 35$, from which $R' = 88.75$ tons = 177,500 pounds.

In like manner to find R, the centre of moments is taken at R'.

Then $R \times 40 = 30 \times 5 + 50 \times 10 + 60 \times 25 + 20 \times 35$, from which $R = 71.25$ tons or 142,500 pounds. As a check, the sum of R' and R is seen to be 160 tons, which is the sum of the four loads.

The thickness of web is then determined by the greatest shear.

$$\text{At } R', \ t = \frac{177500}{36 + 6000} = .821, \text{ say } \tfrac{7}{8} \text{ inch.}$$

Having two webs, each will be $\tfrac{7}{16}''$ thick.

Stiffeners.—To determine whether we require stiffeners:
Safe resistance to buckling

$$= \frac{10000}{1 + \dfrac{36 \times 36}{3000 \times \tfrac{7}{16} \times \tfrac{7}{16}}} = 3069 \text{ pounds per square inch.}$$

Having adopted 6000 lbs. per square inch for safe shearing, the webs will require to be stiffened at the bearings, also under each concentrated load and between W_3 and W_1, W_4 and R' support, two stiffeners between W_2 and W_1, and one between W_3 and R support. Between loads W_2 and W_3 there is very little shear (see diagram Fig.   ).

By riveting $4'' \times 4'' \times \tfrac{7}{16}''$ angles on the outside of webs, we have for the thickness of each web $\tfrac{7}{16}'' + \tfrac{7}{16}'' = \tfrac{14}{16}''$.

Then the formula becomes

$$\frac{10000}{1 + \dfrac{36 \times 36}{3000 \times \tfrac{14}{16} \times \tfrac{14}{16}}} = 6392 \text{ lbs. per square inch.}$$

Rivets.—Using $\tfrac{7}{8}$-inch-diameter rivets, the safe shear per square inch from table equals 4510 lbs.

EXAMPLE VII.

The maximum bending moments at h:

$$M = 868.75 \text{ ton-feet.}$$

Horizontal strain $= \dfrac{868.75}{3} = 289.58$ tons or $579,160$ pounds.

This divided by $4510 = 128$ rivets, to be placed a distance of 180 in one or 360 inches in both webs, spaced about $2\frac{13}{16}$ inches from h to R support.

The maximum bending moment at i:

$$M = 781.25 \text{ ton-feet.}$$

Horizontal strain $= \dfrac{781.25}{3} = 260.41$ tons or $520,820$ pounds.

This divided by $4510 = 115$ rivets, to be placed a distace of 180 in one or 360 inches in both webs, spaced about $3\frac{1}{8}$ inches from i to R' support.

The rivets between h and i to be spaced 6 inches centres.

Flange Plates Reduced in Area towards the Supports.—Draw the Diagram Fig. 43 equal and similar to the polygon Fig. 42, $a, b, c, d, e,$ and f representing the bending moments due to the sum of all the loads at the points $g, h, i,$ and k.

Draw the rectangle $RCER'$. Then from F place the scale at any angle, as at Fe, until it measures 48.2 square inches of the top flange. For two angles set off 5.37 square inches each at a and b; three plates $20'' \times \frac{5}{8}''$ or 12.5 square inches each at $c, d,$ and e. Horizontal lines drawn from $a, b, c, d,$ and e to the polygon and carried down to base line RR' will give the position of plates in each flange.

The angles and adjoining plates to extend the full length of girder.

The plates to extend over the calculated lengths equal to four cross-lines of rivets.

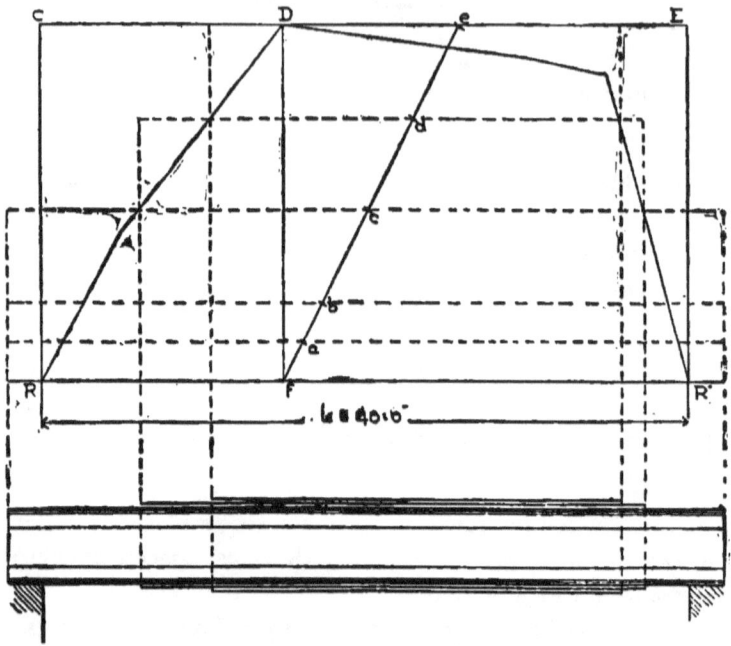

Fig. 43.

Graphical Representation of the Bending Moments and the Shearing Forces for Four Concentrated Loads.—We have in this example a girder of 40 feet span to sustain concentrated loads of 20 tons 5 feet and 60 tons 15 feet from R support, 50 tons 10 feet and 30 tons 5 feet from R' support.

Set the given loads W_1, W_2, W_3, and W_4, Fig. 44, off in succession along a line PP': W_1 of 20 tons at 1, W_2 of 60 tons at 2, W_3 of 50 tons at 3, and W_4 of 30 tons at 4. The line PP' is thus the polygon of the forces W_1, W_2, W_3, and W_4, and PP', its closing line, is their resultant. Take any point O as pole, *equal to* ten *units of the scale adopted*, and draw the radii OP, $O1$, $O2$, $O3$, $O4$. Then describe the funicular polygon

EXAMPLE VII.

abcdef, by drawing *ab* parallel to *OP*, terminating in *W*, produced; *bc* parallel to *O*1, *cd* parallel to *O*2, *de* parallel to *O*3, *ef* parallel to *O*4, terminating in the prolongation of *R'* down-

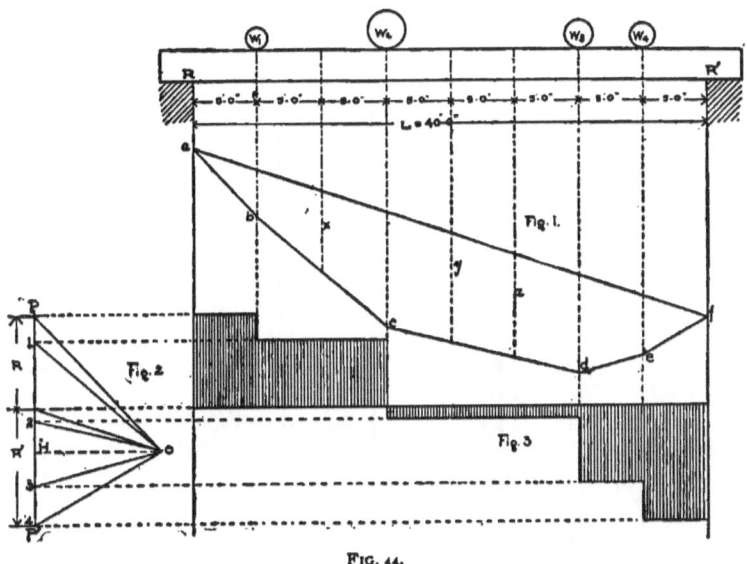

FIG. 44.

wards. The funicular polygon is now closed by the line *fa*, and a line *OS* is drawn through the pole *O* parallel to *fa*. Then as a condition of equilibrium

the reaction at $R = PS$, and at $R' = SP'$.

The bending moments at any point in the girder can be measured by the ordinates of the funicular polygon multiplied by the pole distance *OH*.

The shearing forces on the webs to be measured from the hatched figure.

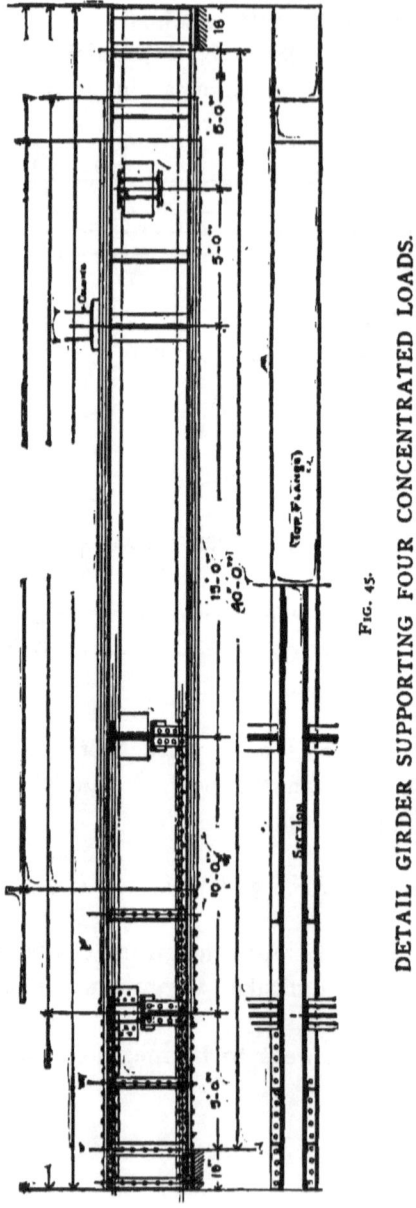

Fig. 45.

DETAIL GIRDER SUPPORTING FOUR CONCENTRATED LOADS.

Span, 40 feet. Depth, 3' 0".

Loads: 20 tons, 5 feet from left support.
 60 "   15 feet from left support.
 50 "   10 feet from right support.
 30 "   5 feet from right support.

Top flange: angles 5" × 3¼" × 1⅛" thick.
 3 plates 20" × ⅝" thick.

Bottom flange: angles 5" × 3¼" × 1⅛" thick.
 2 plates 20" × ⅝" thick.
 1 plate 20" × ⅜" "

Two webs, 1⅛" × 3' 0". Stiffeners, 4" × 4" × 1⅛" angles. Rivets, ⅞ inch diameter.

Example VIII.

STEEL GIRDER SUPPORTING FIVE CONCENTRATED LOADS.

In this example the calculations for the shear on the webs and the transverse strains at any cross-section of the girder will be purely graphical.

Example: What metal area would be required in the flanges of a *steel box girder* of 50 feet span, 4 feet in depth, to sustain the following concentrated loads: 30 tons 10 feet, 40 tons 15 feet, 60 tons 20 feet from left support, 100 tons 20 feet and 50 tons 10 feet from right support?

7 tons unit strain per square inch for the flanges, and 7000 pounds safe shear per square inch on the webs.

Draw the diagram representing the girder, Fig 47, to a scale of $\frac{1}{4}$ of an inch per foot, with two supports and loaded with W_1, W_2, W_3, W_4, and W_5, representing each concentrated load respectively.

Set the given loads off in succession along the line PP' by same scale, equal to 10 tons per foot: W_1 of 30 tons 3 feet at 1; W_2 of 40 tons 4 feet at 2; W_3 of 60 tons 6 feet at 3; W_4 of 100 tons 10 feet at 4; W_5 of 50 tons 5 feet at 5. The line PP' is then the polygon of the forces $W_1 \ldots W_5$, and $P'P$, its closing line, their resultant.

Take any point O as pole, say 10 feet distant from PP', and draw the radii OP, $O1$, $O2$, $O3$, $O4$, and $O5$. Then describe the funicular polygon $abcdefg$, by drawing ab parallel to OP, terminating in W_1 produced; bc parallel to $O1$, terminating in the prolongation of $W_2 \ldots$; and finally fg parallel to $O5$, terminating in the prolongation downwards of R'.

Close the funicular polygon by the line ga, and draw through the pole O a line OS parallel to ga.

Then the reaction at R will equal the distance measured from S to P, and the reaction at R' the distance from S to P'; or, the shearing forces are equal to the distances of the various points of the funicular polygon of forces from S. Accordingly the shearing forces have been taken from the polygon of forces

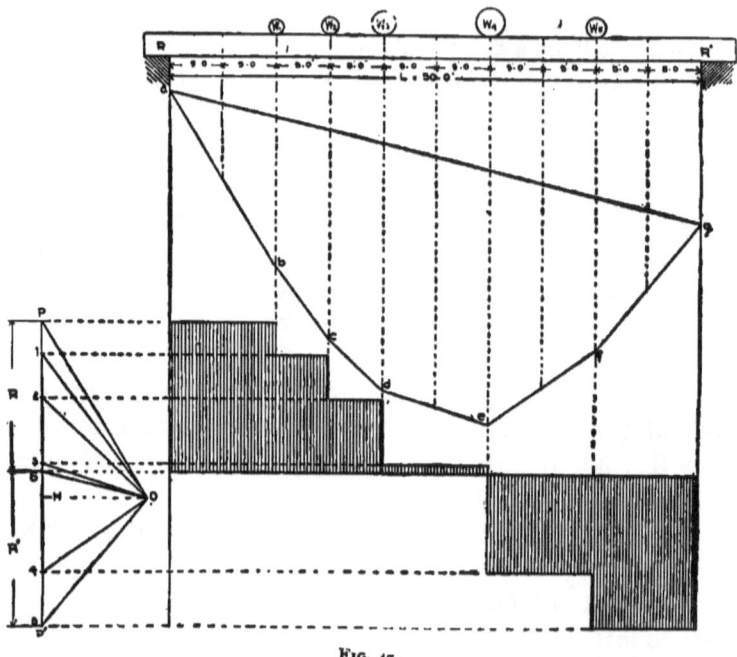

Fig. 47.

and used as segments of PP', to which they correspond; thus the hatched figure is obtained which is termed the "shearing-force diagram."

The shear on the web by scale:

At R = 13 feet $9\frac{5}{8}$ inches,* or 13.8 × 10 = 138 tons.
At R' = 14 " $2\frac{3}{8}$ " " 14.2 × 10 = 142 "

---

* To facilitate the calculation it would be well to adopt a scale in tenths instead of in feet and inches.

Then the thickness of webs at R,

$$T = \frac{276000}{48 \times 7000} = .821, \text{ say } \tfrac{7}{8} \text{ inch.}$$

The girder having two webs, each will be $\tfrac{7}{16}$ of an inch thick.

$$\text{At } R', \ T = \frac{184000}{48 \times 7000} = .845, \text{ say } \tfrac{7}{8} \text{ inch.}$$

Each web will therefore, require to be $\tfrac{7}{16}$ of an inch thick from end to end, and 48 inches in depth.

To repeat:

The shear under load $W_1 = \ldots\ldots\ldots\ldots\ldots\ldots 138$ tons.
" " " " $W_2 = 138 - 30 \qquad = 108$ "
" " " " $W_3 = 138 - (30 + 40) = 68$ "
" " " " $W_4 = 142 - 50 \qquad = 92$ "
" " " " $W_5 = \ldots\ldots\ldots\ldots\ldots\ldots 142$ "

From the above it will be noticed that the thickness of webs would vary from each concentrated load towards the supports; if they were proportioned accordingly, it is doubtful whether the saving in iron would compensate for the additional splicing, punching, and riveting; in very large girders there would be, but for girders 4 feet in depth and under it is hardly practicable.

Determination of the Bending Moments.—The maximum bending moment in the girder is at the greatest load, W_4, and the distance by scale from e in the funicular polygon to the closing line ga measures 23 feet $4\tfrac{13}{16}$ inches, or 23' 4''. This multiplied by 10 (as we adopted a scale of 10 tons per foot), and again by the pole distance, 10, $(OH_4) = 2340$ ton-feet.

$$\text{Area of flange} = \frac{2340}{4 \times 7} = 83.57 \text{ square inches at } W_4.$$

At W_1, M measured from b to the line $ga = 13$ feet $9\frac{5}{8}$ inches, or $13.8 \times 10 \times 10 = 1380$ ton-feet.

$$A = \frac{1380}{4 \times 7} = 49 \text{ square inches.}$$

At W_2, M measured from $C = 19$ feet $2\frac{3}{8}$ inches, or $19.2 \times 10 \times 10 = 1920$ ton-feet.

$$A = \frac{1920}{4 \times 7} = 68.57 \text{ square inches.}$$

At W_3, M measured from $D = 22$ feet $7\frac{3}{16}$ inches, or $22.6 \times 10 \times 10 = 2260$ ton-feet.

$$A = \frac{2260}{4 \times 7} = 80.7 \text{ square inches.}$$

At W_4, M measured from $f = 14$ feet $2\frac{3}{8}$ inches, or $14.2 \times 10 \times 10 = 1420$ ton-feet.

$$A = \frac{1420}{4 \times 7} = 50.7 \text{ square inches.}$$

Construction of Flanges.—To make up the 83.57 square inches in the top flange at e, we construct the section as follows:

Top flange $= 6'' \times 6'' \times 1\frac{3}{8}'' = 18.12$ square inches.
2 plates $24'' \times \frac{3}{4}'' = 36.00$ " "
2 " $24'' \times \frac{5}{8}'' = 30.00$ " "

Total, 84.12 " "

For the bottom flange deduct two rivet-holes in the horizontal leg of the angles connecting the flange and two holes in the vertical leg connecting the webs, then using $\frac{7}{8}$-diameter rivets and allowing $\frac{1}{8}''$ more, we have for one hole in the flange $1\frac{3}{8}'' + \frac{3}{4}'' + \frac{3}{4}'' + \frac{5}{8}'' + \frac{5}{8}'' \times 1'' = 3\frac{9}{16}''$, for two holes $3\frac{9}{16}'' \times 2'' = 7\frac{1}{8}$ inches.

EXAMPLE VIII.

For one hole in the vertical leg of the angle $\frac{3}{4} \times 1 = \frac{3}{4}$; for two holes $\frac{3}{4} \times 2 = 1\frac{1}{2}$ square inches. Then for the bottom flange $83.57 + 7.125 + 1.5 = 92.195$ square inches.

Bottom flange = 2 angles $6'' \times 6'' \times \frac{13}{16}'' = 18.12$ sq. inches.
              1 plate $24'' \times \frac{7}{8}''$ = 21.00 "   "
              3 plates $24'' \times \frac{3}{4}''$ = 54.00 "   "

                                  Total,   93.12 "   "

Stiffeners.—To determine whether we require stiffeners:
Safe resistance to buckling for steel

$$= \frac{11000}{1 + \frac{48 \times 48}{3000 \times \frac{7}{16} \times \frac{7}{16}}} = 2200 \text{ lbs. per square inch.}$$

Having adopted 7000 pounds per square inch for shearing, the webs will require to be stiffened at the bearings, under each concentrated load; one stiffener between R and W_1, one stiffener between W_1 and W_2, and two between W_2 and R' support.

By riveting $5 \times 5 \times \frac{3}{4}$ angles on the outside of the webs at the above points, we have for the thickness of each web $\frac{3}{4} + \frac{7}{16} = \frac{19}{16}$ inches.

Then the formula

$$= \frac{11000}{1 + \frac{48 \times 48}{3000 \times \frac{19}{16} \times \frac{19}{16}}} = 7448 \text{ lbs. per square inch.}$$

Rivets.—Using $\frac{7}{8}$-inch-diameter rivets, the safe shear per square inch equals 4510 pounds.

The maximum bending moment at e:

$$M = 2340 \text{ ton-feet.}$$

Horizontal strain $= \dfrac{2340}{4} = 585$ tons or 117,000 pounds.

This divided by 4510 = 259 rivets, to be placed a distance of 240 inches in one or 480 inches in both webs, spaced about 1⅞ inches (staggered) from e to R' support.

From E to R support the 259 rivets to be placed a distance of 360 inches in one or 720 inches in both webs, spaced about 2 13/16 inches. On account of the greatest horizontal increments of strain in the web at the supports, the rivets should be spaced closer as the ends are approached.

Area of Flanges reduced towards the Supports.—Construct the diagram Fig. 48 by drawing the polygon $RBCDGR'$ similar to the funicular polygon, Fig. 47.

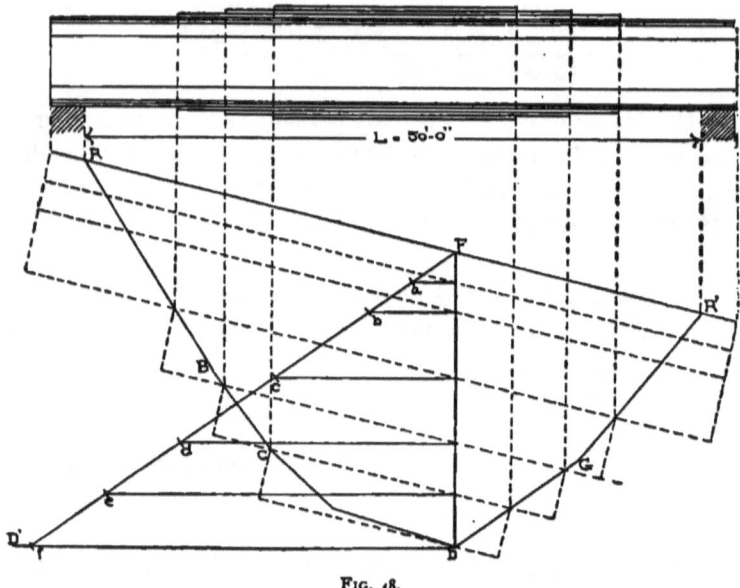

FIG. 48.

From a point on the closing line RR' directly under the maximum bending moment at F draw FD. At F place the scale at any angle, as at Fg, on a horizontal line DD', until it measures 83.57 square inches of the top flange. For two angles set off 8.61 square inches each at a and b; two plates 24 × ⅞, or

18.36 square inches, each at c and d; two plates $24 \times \frac{5}{8}$, or 15.3 square inches, each at e and f. The additional inch in the top plate extends beyond the line $D'D$. Horizontal lines drawn from $ab \ldots f$ to FD, then to the polygon parallel to $RBCDGR'$, and carried up to the base line RR', will give the position of the plates in both flanges.

The angles and adjoining plates to extend the full length of girder.

The other plates of the flanges to extend longer than the calculated distance, as previously explained.

A Cantilever Girder.—A simple girder is a girder resting upon two supports; a *cantilever* girder rests upon one support, in its middle, or the portion of any girder projecting out of a wall or beyond a support.

We have so far considered the effect of loads on girders supported at each end.

The bending moment of a girder resting on one support, as a cantilever, is equal to the weight multiplied by the distance from the weight to the support, or

$$M = WL.$$

The bending moment at *any point* in the girder when the weight is at the end is equal to the weight multiplied by the distance from the support *minus* the distance from the support to point desired, or

$$M = W(L - x).$$

Suppose we have the cantilever Fig. 49, loaded at the end with W. Then will the bending moment at any section, as at x, be obtained by multiplying W by bx; that at RF being WL.

If now we lay off FD to scale to represent this, and join D with the end of girder, then will any ordinate, as x or y, represent (to the same scale) the bending moment at a section x or y.

The strains in top and bottom flanges are reversed to those in a simple girder supported at both ends—the bottom flange is

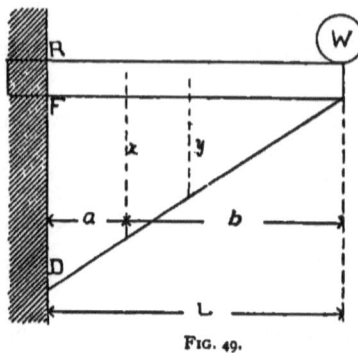

Fig. 49.

in compression, while the top is in tension; but the calculations for the rivets, buckling and shearing of the webs are the same. The shear on the web will be equal to the total load, or

$$S = W.$$

Girder fixed at one End supporting a Uniformly Distributed Load.—If we have a uniformly distributed load, we would have for the line corresponding to FD a curve (concave), or

$$M = \frac{WL}{2}.$$

Then any ordinate, as x or y, drawn to the curved line will represent the bending moments at the corresponding point of the girder.

The shearing stress will equal the total load the same as before.

EXAMPLE VIII.

Girder fixed at one End and supporting Two or More Loads.—When we have more than one load on the girder we must draw the line D to W for each load separately, and then find the actual bending moment at any point by taking the sum of the ordinates (drawn from that point) of each of these separate straight lines or curves.

If we then draw a new curve whose ordinates are these sums, we shall have the actual bending moments for the girder as loaded.

The Relative Strength of Cantilever and Simple Beams.—The following table exhibits the most important results relating to the relative strength of cantilever and simple girders:

Uniform Cross-section.	Maximum Vertical Shear.	Maximum Bending Moment.	Maximum Deflection.	Relative Strength.
Cantilever, load at end..........	W	WL	$\frac{1}{3}\frac{WL^3}{EI}$	1
Cantilever uniformly loaded.....	W	$\frac{1}{2}WL$	$\frac{1}{8}\frac{WL^3}{EI}$	$2\frac{2}{3}$
Simple girder, load at centre....	$\frac{1}{2}W$	$\frac{1}{4}WL$	$\frac{1}{48}\frac{WL^3}{EI}$	16
Simple girder uniformly loaded..	$\frac{1}{2}W$	$\frac{1}{8}WL$	$\frac{5}{384}\frac{WL^3}{EI}$	$25\frac{3}{5}$

$W =$ load in inches; $L =$ span in inches; $E =$ modulus of elasticity in pounds-inch; $I =$ moment of inertia of cross-section in inches.

Experiments on riveted girders have given moduli of elasticity considerably lower than for solid sections, as I beams. Owing, however, to imperfections in riveting, etc., the compound section will deflect much more.

Therefore about 15 per cent less than that adopted for the solid section should be used, say 22,000,000 for wrought-iron and 24,000,000 for steel, 27,000,000 and 29,000,000 being the average for solid sections.

COMPOUND RIVETED GIRDERS.

The Moment of Inertia for Rectangular Sections, such Compound Riveted Girders, as Fig. 50 and Fig. 51.

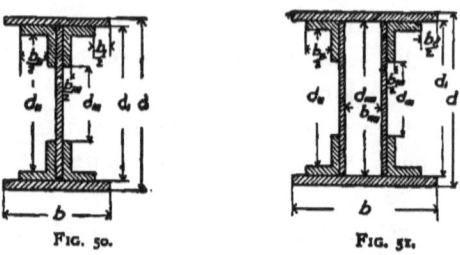

Fig. 50.   Fig. 51.

Fig. 50, $I = \dfrac{bd^3 - b_{/}d_{/}^3 - b_{//}d_{//}^3 - b_{///}d_{///}^3}{12}$.

Fig. 51, $I = \dfrac{bd^3 - b_{/}d_{/}^3 - b_{//}d_{//}^3 - b_{///}d_{///}^3 - b_{////}d_{////}^3}{12}$.

N.B.—The value of $\dfrac{b_{/}}{2}$, $\dfrac{b_{//}}{2}$, $\dfrac{b_{///}}{2}$, and $\dfrac{b_{////}}{2}$ represents one side only; for instance, if $\dfrac{b_{/}}{2} = \tfrac{3}{4}$ of an inch, $b_{/} = 1\tfrac{1}{2}$ inch. The same applies to $\dfrac{d_{/}}{2}$.

PART IV.

TABLES.

PART IV.

TABLES.

AVERAGE WEIGHT, IN POUNDS, OF A CUBIC FOOT OF VARIOUS SUBSTANCES

Aluminum,	162
Anthracite, solid, of Pennsylvania,	93
" broken, loose,	54
" heaped bushel, loose,	(80)
Ash, American white, dry,	38
Asphaltum,	87
Brass, (Copper and Zinc,) cast,	504
" rolled,	524
Brick, best pressed,	150
" common hard,	125
" soft, inferior,	100
Brickwork, pressed brick,	140
" ordinary,	112
Cement, hydraulic, ground, loose, American, Rosendale,	56
" " " " " Louisville,	50
" " " " English, Portland,	90
Cherry, dry,	42
Chestnut, dry,	41
Clay, potter's, dry,	119
" in lump, loose,	63
Coal, bituminous, solid,	84
" " broken, loose,	49
" " heaped bushel, loose,	(74)
Coke, loose, of good coal,	62
" " heaped bushel,	(40)
Copper, cast,	542
" rolled,	548
Earth, common loam, dry, loose,	76
" " " " moderately rammed,	95
" as a soft flowing mud,	108
Ebony, dry,	76
Elm, dry,	35
Flint,	162
Glass, common window,	157
Gneiss, common,	168

PART IV.

TABLES.

AVERAGE WEIGHT, IN POUNDS, OF A CUBIC FOOT OF VARIOUS SUBSTANCES

Aluminum,	162
Anthracite, solid, of Pennsylvania,	93
" broken, loose,	54
" heaped bushel, loose,	(80)
Ash, American white, dry,	38
Asphaltum,	87
Brass, (Copper and Zinc,) cast,	504
" rolled,	524
Brick, best pressed,	150
" common hard,	125
" soft, inferior,	100
Brickwork, pressed brick,	140
" ordinary,	112
Cement, hydraulic, ground, loose, American, Rosendale,	56
" " " " " Louisville,	50
" " " " English, Portland,	90
Cherry, dry,	42
Chestnut, dry,	41
Clay, potter's, dry,	119
" in lump, loose,	63
Coal, bituminous, solid,	84
" " broken, loose,	49
" " heaped bushel, loose,	(74)
Coke, loose, of good coal,	62
" " heaped bushel,	(40)
Copper, cast,	542
" rolled,	548
Earth, common loam, dry, loose,	76
" " " " moderately rammed,	95
" as a soft flowing mud,	108
Ebony, dry,	76
Elm, dry,	35
Flint,	162
Glass, common window,	157
Gneiss, common,	168

WEIGHT OF A CUBIC FOOT OF VARIOUS SUBSTANCES.

Gold, cast, pure, or 24 carat,	1204
" pure, hammered,	1217
Granite,	170
Gravel, about the same as sand, which see.	
Gypsum (plaster of paris),	142
Hemlock dry,	25
Hickory, dry,	53
Hornblende, black,	203
Ice,	58.7
Iron, cast,	450
" wrought, purest,	485
" " average,	480
Ivory,	114
Lead,	711
Lignum Vitæ, dry,	83
Lime, quick, ground, loose, or in small lumps,	53
" " " " thoroughly shaken,	75
" " " " per struck bushel,	(66)
Limestones and Marbles,	168
" " " loose, in irregular fragments,	96
Magnesium,	109
Mahogany, Spanish, dry,	53
" Honduras, dry,	35
Maple, dry,	49
Marbles, see Limestones.	
Masonry, of granite or limestone, well dressed,	165
" " mortar rubble,	154
" " dry " (well scabbled,)	138
" " sandstone, well dressed,	144
Mercury, at 32° Fahrenheit,	849
Mica,	183
Mortar, hardened,	103
Mud, dry, close,	80 to 110
" wet, fluid, maximum,	120
Oak, live, dry,	59
" white, dry,	50
" other kinds,	32 to 45
Petroleum,	55
Pine, white, dry,	25
" yellow, Northern,	34
" " Southern,	45
Platinum,	1342
Quartz, common, pure,	165
Rosin,	69
Salt, coarse, Syracuse, N. Y.,	45
" Liverpool, fine, for table use,	49

Sand, of pure quartz, dry, loose,	90 to 106
" well shaken,	99 to 117
" perfectly wet,	120 to 140
Sandstones, fit for building,	151
Shales, red or black,	162
Silver,	655
Slate,	175
Snow, freshly fallen,	5 to 12
" moistened and compacted by rain,	15 to 50
Spruce, dry,	25
Steel,	490
Sulphur,	125
Sycamore, dry,	37
Tar,	62
Tin, cast,	459
Turf or Peat, dry, unpressed,	20 to 30
Walnut, black, dry,	38
Water, pure rain or distilled, at 60° Fahrenheit,	$62\tfrac{1}{2}$
" sea,	64
Wax, bees,	60.5
Zinc or Spelter,	437

WEIGHT OF 100 RIVETS IN POUNDS.

Length of Rivet in Inches under Head.	Diameter of Rivet in Inches.					
	⅜	½	⅝	¾	⅞	1
1¼	5.4	12.5	21.2	28.0	42.5	64.6
1⅜	5.9	13.1	22.4	29.5	44.6	67.3
1½	6.3	13.7	23.5	31.0	46.7	69.9
1⅝	6.7	14.4	24.7	32.7	48.9	72.8
1¾	7.0	15.1	26.0	34.2	51.0	75.0
1⅞	7.3	15.8	27.1	35.6	53.3	77.8
2	7.6	16.5	28.3	37.0	55.2	81.3
2⅛	7.9	17.2	29.6	38.4	57.5	84.1
2¼	8.3	17.8	31.0	39.8	59.5	86.9
2⅜	8.8	18.4	32.1	41.5	61.7	89.5
2½	9.1	19.1	33.2	43.2	63.9	92.2
2⅝	9.5	19.8	34.4	44.8	66.0	94.8
2¾	9.8	20.5	35.4	46.1	68.2	97.3
2⅞	10.2	21.2	36.1	47.7	70.1	100.0
3	10.6	21.9	37.0	49.0	72.1	102.5
3⅛	11.0	22.7	38.2	50.6	74.0	105.1
3¼	11.3	23.4	39.1	52.1	76.2	107.8
3⅜	11.7	24.0	40.2	53.7	78.5	110.4
3½	12.1	24.7	41.0	55.2	80.2	112.9
3⅝	12.5	25.3	42.0	56.7	82.4	115.5
3¾	12.8	26.0	42.9	58.1	84.3	118.0
3⅞	13.2	26.6	44.1	60.0	86.5	120.6
4	13.6	27.2	45.1	61.5	88.7	123.2
4⅛	14.0	28.0	46.2	63.2	91.0	125.7
4¼	14.4	28.9	47.1	65.1	93.4	128.3
4⅜	14.9	29.5	48.0	66.6	95.1	131.0
4½	15.3	30.2	48.9	68.0	97.3	133.6
4⅝	15.7	30.9	49.8	69.2	99.5	136.2
4¾	16.1	31.6	51.0	70.9	101.1	138.8
4⅞	16.5	32.2	52.1	72.5	103.4	141.3
5	17.0	32.9	53.3	74.2	105.2	144.0
5¼	17.6	33.6	55.6	77.2	109.8	150.0
5½	18.2	35.1	56.8	80.3	114.1	155.7
5¾	18.9	36.6	58.0	83.2	118.0	161.0
6	19.7	37.7	59.9	86.1	122.7	166.1

WEIGHT OF TWO (2) RIVET-HEADS IN POUNDS.

	⅜	½	⅝	¾	⅞	1
Before driving	.036	.114	.218	.268	.444	.76
After driving	.031	.080	.160	.260	.440	.64

WEIGHT OF BODY PER INCH OF LENGTH.

	⅜	½	⅝	¾	⅞	1
Before driving	.031	.054	.085	.123	.167	.218

DECIMAL EQUIVALENTS FOR FRACTIONS OF A FOOT.

1/16	.0052	3 1/16	.2552	6 1/16	.5052	9 1/16	.7552
1/8	.0104	3 1/8	.2604	6 1/8	.5104	9 1/8	.7604
3/16	.0156	3 3/16	.2656	6 3/16	.5156	9 3/16	.7656
1/4	.0208	3 1/4	.2708	6 1/4	.5208	9 1/4	.7708
5/16	.0260	3 5/16	.2760	6 5/16	.5260	9 5/16	.7760
3/8	.0312	3 3/8	.2812	6 3/8	.5312	9 3/8	.7812
7/16	.0364	3 7/16	.2865	6 7/16	.5364	9 7/16	.7865
1/2	.0417	3 1/2	.2917	6 1/2	.5411	9 1/2	.7917
9/16	.0469	3 9/16	.2969	6 9/16	.5469	9 9/16	.7969
5/8	.0521	3 5/8	.3021	6 5/8	.5521	9 5/8	.8021
11/16	.0573	3 11/16	.3073	6 11/16	.5573	9 11/16	.8073
3/4	.0625	3 3/4	.3125	6 3/4	.5625	9 3/4	.8125
13/16	.0677	3 13/16	.3177	6 13/16	.5677	9 13/16	.8177
7/8	.0729	3 7/8	.3229	6 7/8	.5729	9 7/8	.8229
15/16	.0781	3 15/16	.3281	6 15/16	.5781	9 15/16	.8281
1	.0833	4	.3333	7	.5833	10	.8333
1 1/16	.0885	4 1/16	.3385	7 1/16	.5885	10 1/16	.8385
1 1/8	.0937	4 1/8	.3437	7 1/8	.5937	10 1/8	.8437
1 3/16	.0990	4 3/16	.3490	7 3/16	.5990	10 3/16	.8490
1 1/4	.1042	4 1/4	.3542	7 1/4	.6042	10 1/4	.8542
1 5/16	.1094	4 5/16	.3594	7 5/16	.6094	10 5/16	.8594
1 3/8	.1146	4 3/8	.3646	7 3/8	.6146	10 3/8	.8646
1 7/16	.1198	4 7/16	.3698	7 7/16	.6198	10 7/16	.8698
1 1/2	.1250	4 1/2	.3750	7 1/2	.6250	10 1/2	.8750
1 9/16	.1302	4 9/16	.3802	7 9/16	.6302	10 9/16	.8802
1 5/8	.1354	4 5/8	.3854	7 5/8	.6354	10 5/8	.8854
1 11/16	.1406	4 11/16	.3906	7 11/16	.6406	10 11/16	.8906
1 3/4	.1458	4 3/4	.3958	7 3/4	.6458	10 3/4	.8958
1 13/16	.1510	4 13/16	.4010	7 13/16	.6510	10 13/16	.9010
1 7/8	.1562	4 7/8	.4062	7 7/8	.6562	10 7/8	.9062
1 15/16	.1615	4 15/16	.4114	7 15/16	.6615	10 15/16	.9115
2	.1667	5	.4167	8	.6667	11	.9167
2 1/16	.1719	5 1/16	.4219	8 1/16	.6719	11 1/16	.9219
2 1/8	.1771	5 1/8	.4271	8 1/8	.6771	11 1/8	.9271
2 3/16	.1823	5 3/16	.4323	8 3/16	.6823	11 3/16	.9323
2 1/4	.1875	5 1/4	.4375	8 1/4	.6875	11 1/4	.9375
2 5/16	.1927	5 5/16	.4427	8 5/16	.6927	11 5/16	.9427
2 3/8	.1979	5 3/8	.4479	8 3/8	.6979	11 3/8	.9479
2 7/16	.2031	5 7/16	.4531	8 7/16	.7031	11 7/16	.9531
2 1/2	.2083	5 1/2	.4583	8 1/2	.7083	11 1/2	.9583
2 9/16	.2135	5 9/16	.4635	8 9/16	.7135	11 9/16	.9635
2 5/8	.2187	5 5/8	.4688	8 5/8	.7187	11 5/8	.9687
2 11/16	.2240	5 11/16	.4740	8 11/16	.7240	11 11/16	.9740
2 3/4	.2292	5 3/4	.4792	8 3/4	.7292	11 3/4	.9792
2 13/16	.2344	5 13/16	.4844	8 13/16	.7344	11 13/16	.9844
2 7/8	.2395	5 7/8	.4896	8 7/8	.7396	11 7/8	.9896
2 15/16	.2448	5 15/16	.4948	8 15/16	.7448	11 15/16	.9948
3	.2500	6	.5000	9	.7500	12	1.000

NUMBER OF U. S. GALLONS (231 CUBIC INCHES) CONTAINED IN CIRCULAR TANKS.

Depth in feet.	1	2	3	4	5	6	7	8	9	10
Dia.	Gals.	Gals.	Gals.	Gals.	Gals.	Gals.	Gals.	Gals.	Gals.	Gals.
In.										
20	16.32	32.64	48.96	65.28	81.60	97.92	114.24	130.56	146.88	163.2
24	23.50	47.00	70.50	94.00	117.50	141.00	164.50	188.00	211.50	235.0
26	27.58	55.16	82.74	110.32	137.90	165.48	193.06	220.64	248.22	275.8
28	31.99	63.98	95.97	127.96	159.95	191.94	223.93	255.92	288.91	319.9
30	36.72	73.44	110.16	146.88	183.60	220.32	257.04	293.76	330.48	367.2
36	52.88	105.76	158.64	211.52	264.40	317.28	370.16	423.04	475.92	528.8
42	71.96	143.92	215.88	287.84	359.80	431.76	503.72	575.68	647.64	719.6
45	82.62	165.24	247.86	330.48	413.10	495.72	578.34	660.96	743.58	826.2
48	94.02	188.04	282.06	376.08	470.10	564.12	658.14	752.16	846.18	940.2
50	102.00	204.00	306.00	408.00	510.00	612.00	714.00	816.00	918.00	1020.0
54	119.00	238.00	357.00	476.00	595.00	714.00	833.00	952.00	1071.00	1190.0
60	146.90	293.80	440.70	587.60	734.50	881.40	1028.30	1175.20	1322.10	1469.0
66	177.70	355.40	533.10	710.80	888.50	1066.20	1243.90	1421.60	1599.20	1777.0
72	211.50	423.00	634.50	846.00	1057.50	1269.00	1480.50	1692.00	1903.50	2115.0
84	287.80	575.60	863.40	1151.20	1439.00	1726.80	2014.60	2302.40	2590.20	2878.0

DECIMAL EQUIVALENTS FOR FRACTIONS OF AN INCH.

Fraction.	Decimal.	Fraction.	Decimal.	Fraction.	Decimal.	Fraction.	Decimal.
1/64	.015625	17/64	.265625	33/64	.515625	49/64	.765625
1/32	.03125	9/32	.28125	17/32	.53125	25/32	.78125
3/64	.046875	19/64	.296875	35/64	.546875	51/64	.796875
1/16	.0625	5/16	.3125	9/16	.5625	13/16	.8125
5/64	.078125	21/64	.328125	37/64	.578125	53/64	.828125
3/32	.09375	11/32	.34375	19/32	.59375	27/32	.84375
7/64	.109375	23/64	.359375	39/64	.609375	55/64	.859375
1/8	.125	3/8	.375	5/8	.625	7/8	.875
9/64	.140625	25/64	.390625	41/64	.640625	57/64	.890625
5/32	.15625	13/32	.40625	21/32	.65625	29/32	.90625
11/64	.171875	27/64	.421875	43/64	.671875	59/64	.921875
3/16	.1875	7/16	.4375	11/16	.6875	15/16	.9375
13/64	.203125	29/64	.453125	45/64	.703125	61/64	.953125
7/32	.21875	15/32	.46875	23/32	.71875	31/32	.96875
15/64	.234375	31/64	.484375	47/64	.734375	63/64	.984375
1/4	.25	1/2	.5	3/4	.75		

WEIGHT PER LINEAL FOOT OF CIRCULAR CAST-IRON COLUMNS.

Outside Diam. in inches.	Thickness of Metal in inches.																
	1/2	5/8	3/4	7/8	1	1 1/8	1 1/4	1 3/8	1 1/2	1 5/8	1 3/4	1 7/8	2	2 1/8	2 1/4	2 3/8	2 1/2
3	12.3	14.6	16.6	18.3	19.6	.....	.....	.....	.....	.....	.....	.....	.....	.....	.....	.....	.....
4	17.2	21.0	24.0	27.0	29.5	32.1	33.8	35.4	.....	.....	.....	.....	.....	.....	.....	.....	.....
5	22.1	27.0	31.3	35.5	39.3	43.0	46.0	49.0	51.54	54.1	55.84	57.5	.....	.....	.....	.....	.....
6	27.0	33.0	39.0	44.0	49.1	54.1	58.3	62.4	66.30	69.9	73.02	76.0	78.6	80.84	82.83	.....	.....
7	32.0	39.1	46.0	53.0	59.0	65.1	70.6	76.1	81.00	85.8	90.20	94.3	98.2	101.70	105.00	107.84	110.45
8	36.8	45.3	53.4	61.2	69.1	76.1	83.1	89.5	95.80	101.8	107.40	112.8	117.8	122.60	127.00	131.20	135.00
9	41.7	51.4	61.1	70.0	78.6	87.1	95.1	103.1	110.50	117.7	124.60	131.2	137.5	143.40	149.10	154.50	159.50
10	46.6	57.5	75.5	87.1	88.4	98.0	107.4	116.4	125.20	133.7	142.00	149.6	157.1	164.30	171.20	177.80	184.10
11	51.6	64.0	83.0	96.1	98.2	109.1	120.1	130.1	140.00	149.6	159.00	168.0	176.8	185.20	193.30	201.10	208.60
12	56.5	70.0	90.0	104.2	108.0	120.0	132.1	143.5	154.70	165.6	176.00	186.4	196.4	206.00	215.40	224.40	233.20
13	61.4	76.0	98.1	113.1	118.1	131.2	144.2	157.1	169.40	181.5	193.30	204.8	216.0	226.90	237.50	247.70	257.70
14	66.3	82.1	105.1	121.4	128.1	142.0	156.5	170.4	184.10	197.4	210.50	223.2	235.7	247.70	259.60	271.10	282.30
15	71.2	88.2	112.3	130.1	137.5	153.3	169.4	184.1	198.90	213.4	227.70	241.6	255.3	286.60	281.70	294.40	306.80
16	76.1	94.4	119.7	138.6	147.3	164.3	181.0	197.4	213.50	229.4	244.90	260.0	274.9	289.50	303.70	317.70	331.50
17	81.0	100.5	120.1	139.1	157.1	175.4	193.3	211.0	228.30	245.3	262.00	278.4	294.5	310.30	325.80	341.00	355.90
18	86.0	107.0	127.0	147.0	167.0	186.4	206.0	224.4	243.00	261.3	279.20	296.8	314.2	331.20	348.00	364.30	380.40
19	91.0	113.0	134.4	156.0	177.1	197.5	218.1	238.0	257.70	277.2	296.40	315.2	338.8	352.10	370.00	387.70	405.00
20	96.0	119.0	142.1	164.3	186.6	208.8	230.1	251.5	272.50	293.2	313.60	333.6	353.4	372.90	392.10	411.00	430.00
21	100.6	125.0	149.1	173.1	196.6	219.6	242.4	265.0	287.20	309.0	330.80	352.1	373.1	393.80	414.20	434.30	454.10
22	105.6	131.2	156.5	181.5	206.2	230.6	255.0	278.0	302.00	325.1	348.00	370.5	393.0	414.60	436.30	457.60	478.60
23	110.5	137.3	164.1	190.1	216.1	242.0	267.0	292.0	316.70	341.0	365.10	388.9	412.3	435.50	458.40	481.00	503.20
24	115.4	143.5	171.2	199.1	226.0	253.0	279.2	305.4	331.40	357.0	382.30	407.3	432.0	456.40	480.50	504.20	527.70

(NOTE.—The table is arranged for the weight of plain shaft. For brackets, flanges, etc., calculate the cubical contents and multiply by .26.)

TABLE OF SQUARE CAST-IRON COLUMNS.

WEIGHT OF SQUARE CAST-IRON COLUMNS IN POUNDS PER LINEAL FOOT.

$2a+2b$	\multicolumn{8}{c}{Thickness of Metal in inches.}
	$\frac{5}{8}$	$\frac{3}{4}$	$\frac{7}{8}$	1	$1\frac{1}{8}$	$1\frac{1}{4}$	$1\frac{3}{8}$	$1\frac{3}{4}$	2

$2a+2b$	$\frac{5}{8}$	$\frac{3}{4}$	$\frac{7}{8}$	1	$1\frac{1}{8}$	$1\frac{1}{4}$	$1\frac{3}{8}$	$1\frac{3}{4}$	2
12	18.6	21.1	23.3	25.0	26.4	27.3	28.1	......	......
14	22.5	25.8	28.7	31.3	33.4	35.1	37.5	......	......
16	26.4	30.5	34.2	37.5	40.4	43.0	46.9	49.2	50.0
18	30.3	35.2	39.7	43.8	47.4	50.8	56.3	60.2	62.5
20	34.2	39.8	45.1	50.0	54.5	58.6	65.6	71.1	75.0
22	38.1	44.5	50.6	56.3	61.5	66.4	75.0	82.0	87.5
24	42.0	49.2	56.1	62.5	68.5	74.2	84.4	93.0	100.0
26	45.9	53.9	61.5	68.8	75.6	82.0	93.8	103.9	112.5
28	49.8	58.6	67.0	75.0	82.6	89.8	103.1	114.8	125.0
30	53.7	63.3	72.5	81.3	89.6	97.7	112.5	125.8	137.5
32	57.6	68.0	77.9	87.5	96.7	105.5	121.9	136.7	150.0
34	61.5	72.7	83.4	93.8	103.7	113.3	131.3	147.7	162.5
36	65.4	77.3	88.9	100.0	110.7	121.1	140.6	158.6	175.0
38	69.3	82.0	94.3	106.3	117.8	128.9	150.0	169.5	187.5
40	73.2	86.7	99.8	112.5	124.8	136.7	159.4	180.5	200.0
42	77.1	91.4	105.3	118.8	131.8	144.5	168.8	191.4	212.5
44	81.0	96.1	110.8	125.0	138.8	152.3	178.1	202.3	225.0
46	84.9	100.8	116.2	131.3	145.9	160.2	187.5	213.3	237.5
48	88.8	105.5	121.7	137.5	152.9	168.0	196.9	224.2	250.0
50	92.8	110.2	127.2	143.8	159.9	175.8	206.3	235.2	262.5
52	96.7	114.8	132.6	150.0	167.0	183.6	215.6	246.1	275.0
54	100.6	119.5	138.1	156.3	174.0	191.4	225.0	257.0	287.5
56	104.5	124.2	143.6	162.5	181.0	199.2	234.4	268.0	300.0
58	108.4	128.9	149.0	168.8	188.1	207.0	243.8	278.9	312.5
60	112.3	133.6	154.5	175.0	195.1	214.9	253.2	289.8	325.0
62	116.2	138.3	160.0	181.3	202.1	222.7	262.5	300.8	337.5
64	120.1	143.0	165.4	187.5	209.2	230.5	271.9	311.7	350.0
66	124.0	147.7	170.9	193.8	216.2	238.3	281.3	322.7	362.5
68	127.9	152.3	176.4	200.0	223.2	246.1	290.6	333.6	375.0
70	131.8	157.0	181.8	206.3	230.3	253.9	300.0	344.5	387.5
72	135.7	161.7	187.3	212.5	237.3	261.7	309.4	355.5	400.0
74	139.6	166.4	192.8	218.8	244.3	269.5	318.8	366.4	412.5
76	143.5	171.1	198.3	225.0	251.3	277.3	328.1	377.3	425.0
78	147.4	175.8	203.7	231.3	258.4	285.2	337.5	388.3	437.5
80	151.3	180.5	207.2	237.5	265.4	293.0	346.9	399.2	450.0

* a and b = either side. $2a + 2b$ = number.

EXAMPLE. What is the weight per lineal foot of a $12'' \times 16'' \times 1''$ thick column?

Ans. $2a + 2b = 24 + 36 = 56$. Opposite this number, under 1-inch thick metal, we find 162.5, or weight per lineal foot of a $12'' \times 16'' \times 1''$ thick column.

NOTE.—For flanges, brackets, etc., calculate the cubical contents of same and multiply by .26; cast iron averaging 450 pounds per cubic foot.

WEIGHT PER FOOT OF FLAT IRON.

(For weight per foot of steel add 2 per cent.)

Breadth in inches.	THICKNESS IN FRACTIONS OF INCHES.										
	1/16	1/8	3/16	1/4	5/16	3/8	7/16	1/2	9/16	5/8	11/16
1	.208	.417	.625	.833	1.04	1.25	1.46	1.67	1.88	2.08	2.29
1 1/8	.234	.469	.703	.938	1.17	1.41	1.64	1.87	2.11	2.34	2.58
1 1/4	.260	.521	.781	1.04	1.30	1.56	1.82	2.08	2.34	2.60	2.86
1 3/8	.286	.573	.859	1.15	1.43	1.72	2.01	2.29	2.58	2.86	3.15
1 1/2	.313	.625	.938	1.25	1.56	1.88	2.19	2.50	2.81	3.13	3.44
1 5/8	.339	.677	1.02	1.36	1.69	2.03	2.37	2.71	3.05	3.39	3.73
1 3/4	.365	.729	1.09	1.46	1.82	2.19	2.55	2.92	3.28	3.65	4.01
1 7/8	.391	.781	1.17	1.56	1.95	2.34	2.73	3.12	3.51	3.91	4.30
2	.417	.833	1.25	1.67	2.08	2.50	2.92	3.33	3.75	4.17	4.58
2 1/8	.443	.886	1.33	1.77	2.21	2.65	3.10	3.54	3.98	4.43	4.87
2 1/4	.469	.938	1.41	1.88	2.34	2.81	3.28	3.75	4.22	4.69	5.16
2 3/8	.495	.990	1.48	1.98	2.47	2.97	3.46	3.96	4.46	4.95	5.44
2 1/2	.521	1.04	1.56	2.08	2.60	3.13	3.65	4.17	4.69	5.21	5.73
2 5/8	.547	1.09	1.64	2.19	2.73	3.28	3.83	4.38	4.92	5.47	6.02
2 3/4	.573	1.15	1.72	2.29	2.86	3.44	4.01	4.58	5.16	5.73	6.30
2 7/8	.599	1.20	1.80	2.40	3.00	3.60	4.20	4.79	5.39	5.99	6.59
3	.625	1.25	1.88	2.50	3.13	3.75	4.38	5.00	5.63	6.25	6.88
3 1/4	.677	1.35	2.03	2.71	3.39	4.06	4.74	5.42	6.09	6.77	7.45
3 1/2	.729	1.46	2.19	2.92	3.65	4.38	5.10	5.83	6.56	7.29	8.02
3 3/4	.781	1.56	2.34	3.13	3.91	4.69	5.47	6.25	7.03	7.81	8.59
4	.833	1.67	2.50	3.33	4.17	5.00	5.83	6.67	7.50	8.33	9.17
4 1/4	.885	1.77	2.66	3.54	4.43	5.31	6.20	7.08	7.97	8.85	9.74
4 1/2	.938	1.88	2.81	3.75	4.69	5.63	6.56	7.50	8.44	9.38	10.31
4 3/4	.990	1.98	2.97	3.96	4.95	5.94	6.93	7.92	8.91	9.90	10 89
5	1.042	2.08	3.13	4.17	5.21	6.25	7.29	8.33	9.38	10.42	11.[illegible]
5 1/4	1.09	2.19	3.28	4.38	5.47	6.56	7.66	8.75	9.84	10.94	11.03
5 1/2	1.15	2.29	3.44	4.58	5.73	6.88	8.02	9.17	10.31	11.46	12.60
5 3/4	1.20	2.40	3.59	4.79	5.99	7.19	8.39	9.58	10.78	11.98	13.18
6	1.25	2.50	3.75	5.00	6.25	7.50	8.75	10.00	11.25	12.50	13.75
6 1/4	1.30	2.60	3.91	5.21	6.51	7.81	9.11	10.42	11.72	13.02	14.32
6 1/2	1.35	2.71	4.06	5.42	6.77	8.13	9.48	10.83	12.19	13.54	14.90
6 3/4	1.41	2.81	4.22	5.63	7.03	8.44	9.84	11.25	12.66	14.06	15.47
7	1.46	2.92	4.38	5.83	7.29	8.75	10.21	11.67	13.13	14.58	16.04
7 1/4	1.51	3.02	4.53	6.04	7.55	9.06	10.57	12.08	13.59	15.10	16.61
7 1/2	1.56	3.13	4.69	6.25	7.81	9.38	10.94	12.50	14.06	15.63	17.19
7 3/4	1.61	3.23	4.84	6.46	8.07	9.69	11.30	12.92	14.53	16.15	17.76
8	1.67	3.33	5.00	6.67	8.33	10.00	11.67	13.33	15.00	16.67	18.33
8 1/4	1.72	3.44	5.16	6.88	8.59	10.31	12.03	13.75	15.47	17.19	18.91
8 1/2	1.77	3.54	5.31	7.08	8.85	10.63	12.40	14.17	15.94	17.71	19.48
8 3/4	1.82	3.65	5.47	7.29	9.11	10.94	12.76	14.58	16.41	18.23	20.05
9	1.88	3.75	5.63	7.50	9.38	11.25	13.13	15.00	16.88	18.75	20.63
9 1/4	1.93	3.85	5.78	7.71	9.64	11.56	13.49	15.42	17.34	19.27	21.20
9 1/2	1.98	3.96	5.94	7.92	9.90	11.88	13.85	15.83	17.81	19.79	21.77
9 3/4	2.03	4 06	6.09	8.13	10.16	12.19	14.22	16.25	18.28	20.31	22.34
10	2.08	4.17	6.25	8.33	10.42	12.50	14.58	16.67	18.75	20.83	22.92
10 1/4	2.14	4.27	6.41	8.54	10.68	12.81	14.95	17.08	19.22	21.35	23.49
10 1/2	2.19	4.38	6.56	8.75	10.94	13.13	15.31	17.50	19.69	21.88	24.06
10 3/4	2.24	4.48	6.72	8.96	11.20	13.44	15.68	17.92	20.16	22.40	24.64
11	2 29	4 58	6.88	9.17	11.46	13.75	16.04	18.33	20.63	22.92	25.21
11 1/4	2.34	4.69	7.03	9.38	11.72	14.06	16.41	18.75	21.09	23.44	25.78
11 1/2	2.40	4.79	7.19	9.58	11.98	14.38	16.77	19.17	21.56	23.96	26.35
11 3/4	2.45	4.90	7.34	9.79	12.24	14.29	17.14	19.58	22.03	24.48	26.93
12	2.50	5.00	7.50	10.00	12.50	15.00	17.50	20.00	22.50	25.00	27.50

TABLE OF WEIGHT OF FLAT IRON.

WEIGHT PER FOOT OF FLAT IRON—*Continued.*

(For weight per foot of steel add 2 per cent.)

Breadth in inches	THICKNESS IN FRACTIONS OF INCHES.										
	¾	13/16	⅞	15/16	1	1 1/16	1 ⅛	1 3/16	1 ¼	1 5/16	1 ⅜
1	2.50	2.71	2.92	3.13	3.33	3.54	3.75	3.96	4.17	4.37	4.58
1⅛	2.81	3.05	3.28	3.52	3.75	3.98	4.22	4.45	4.69	4.92	5.16
1¼	3.13	3.39	3.65	3.91	4.17	4.43	4.69	4.95	5.21	5.47	5.73
1⅜	3.44	3.72	4.01	4.30	4.58	4.87	5.16	5.44	5.73	6.02	6.30
1½	3.75	4.06	4.38	4.69	5.00	5.31	5.63	5.94	6.25	6.56	6.88
1⅝	4.06	4.40	4.74	5.08	5.42	5.75	6.09	6.43	6.77	7.11	7.45
1¾	4.38	4.74	5.10	5.47	5.83	6.20	6.56	6.93	7.29	7.66	8.02
1⅞	4.69	5.08	5.47	5.86	6.25	6.64	7.03	7.42	7.81	8.20	8.59
2	5.00	5.42	5.83	6.25	6.67	7.08	7.50	7.92	8.33	8.75	9.17
2⅛	5.31	5.75	6.20	6.64	7.08	7.52	7.97	8.41	8.85	9.30	9.74
2¼	5.63	6.09	6.56	7.03	7.50	7.97	8.44	8.91	9.38	9.84	10.31
2⅜	5.94	6.43	6.93	7.42	7.92	8.41	8.91	9.40	9.90	10.39	10.89
2½	6.25	6.77	7.29	7.81	8.33	8.85	9.38	9.90	10.42	10.94	11.46
2⅝	6.56	7.11	7.66	8.20	8.75	9.30	9.84	10.39	10.94	11.48	12.03
2¾	6.88	7.45	8.02	8.59	9.17	9.74	10.31	10.89	11.46	12.03	12.60
2⅞	7.19	7.79	8.39	8.98	9.58	10.18	10.78	11.38	11.98	12.58	13.18
3	7.50	8.13	8.75	9.38	10.00	10.63	11.25	11.88	12.50	13.13	13.75
3¼	8.13	8.80	9.48	10.16	10.83	11.51	12.19	12.86	13.54	14.22	14.90
3½	8.75	9.48	10.21	10.94	11.67	12.40	13.13	13.85	14.58	15.31	16.04
3¾	9.38	10.16	10.94	11.72	12.50	13.28	14.06	14.84	15.63	16.41	17.19
4	10.00	10.83	11.67	12.50	13.33	14.17	15.00	15.83	16.67	17.50	18.33
4¼	10.63	11.51	12.40	13.28	14.17	15.05	15.94	16.82	17.71	18.59	19.48
4½	11.25	12.19	13.13	14.06	15.00	15.94	16.88	17.81	18.75	19.69	20.63
4¾	11.88	12.86	13.85	14.84	15.83	16.82	17.81	18.80	19.79	20.78	21.77
5	12.50	13.54	14.58	15.63	16.67	17.71	18.75	19.79	20.83	21.88	22.92
5¼	13.13	14.22	15.31	16.41	17.50	18.59	19.69	20.78	21.88	22.97	24.06
5½	13.75	14.90	16.04	17.19	18.33	19.48	20.63	21.77	22.92	24.06	25.21
5¾	14.38	15.57	16.77	17.97	19.17	20.36	21.56	22.76	23.96	25.16	26.35
6	15.00	16.25	17.50	18.75	20.00	21.25	22.50	23.75	25.00	26.25	27.50
6¼	15.63	16.93	18.23	19.53	20.83	22.14	23.44	24.74	26.04	27.34	28.65
6½	16.25	17.60	18.96	20.31	21.67	23.02	24.38	25.73	27.08	28.44	29.79
6¾	16.88	18.28	19.69	21.09	22.50	23.91	25.31	26.72	28.13	29.53	30.94
7	17.50	18.96	20.42	21.88	23.33	24.79	26.25	27.71	29.17	30.62	32.08
7¼	18.13	19.64	21.15	22.66	24.17	25.68	27.19	28.70	30.21	31.72	33.23
7½	18.75	20.31	21.88	23.44	25.00	26.56	28.13	29.69	31.25	32.81	34.38
7¾	19.38	20.99	22.60	24.22	25.83	27.45	29.06	30.68	32.29	33.91	35.52
8	20.00	21.67	23.33	25.00	26.67	28.33	30.00	31.67	33.33	35.00	36.67
8¼	20.63	22.34	24.06	25.78	27.50	29.22	30.94	32.66	34.38	36.09	37.81
8½	21.25	23.02	24.79	26.56	28.33	30.10	31.88	33.65	35.42	37.19	38.96
8¾	21.88	23.70	25.52	27.34	29.17	30.99	32.81	34.64	36.46	38.28	40.10
9	22.50	24.38	26.25	28.13	30.00	31.88	33.75	35.63	37.50	39.38	41.25
9¼	23.13	25.05	26.98	28.91	30.83	32.76	34.69	36.61	38.54	40.47	42.40
9½	23.75	25.73	27.71	29.69	31.67	33.65	35.63	37.60	39.58	41.56	43.54
9¾	24.38	26.41	28.44	30.47	32.50	34.53	36.56	38.59	40.63	42.66	44.69
10	25.00	27.08	29.17	31.25	33.33	35.42	37.50	39.58	41.67	43.75	45.83
10¼	25.62	27.76	29.90	32.03	34.17	36.30	38.44	40.57	42.71	44.84	46.98
10½	26.25	28.44	30.63	32.81	35.00	37.19	39.38	41.56	43.75	45.94	48.13
10¾	26.88	29.11	31.35	33.59	35.83	38.07	40.31	42.55	44.79	47.03	49.27
11	27.50	29.79	32.08	34.38	36.67	38.96	41.25	43.54	45.83	48.13	50.42
11¼	28.13	30.47	32.81	35.16	37.50	39.84	42.19	44.53	46.88	49.22	51.56
11½	28.75	31.15	33.54	35.94	38.33	40.73	43.13	45.52	47.92	50.31	52.71
11¾	29.38	31.82	34.27	36.72	39.17	41.61	44.06	46.51	48.96	51.41	53.85
12	30.00	32.50	35.00	37.50	40.00	42.50	45.00	47.50	50.00	52.50	55.00

TABLE OF SQUARES AND CUBES.

SQUARES AND CUBES, OF NUMBERS FROM 1 TO 440.

No.	Squares.	Cubes.	No.	Squares.	Cubes.
1	1	1	56	31 36	175 616
2	4	8	57	32 49	185 193
3	9	27	58	33 64	195 112
4	16	64	59	34 81	205 379
5	25	125	60	36 00	216 000
6	36	216	61	37 21	226 981
7	49	343	62	38 44	238 328
8	64	512	63	39 69	250 047
9	81	729	64	40 96	262 144
10	1 00	1 000	65	42 25	274 626
11	1 21	1 331	66	43 56	287 496
12	1 44	1 728	67	44 89	300 763
13	1 69	2 197	68	46 24	314 432
14	1 96	2 744	69	47 61	328 509
15	2 25	3 375	70	49 00	343 000
16	2 56	4 096	71	50 41	357 911
17	2 89	4 913	72	51 84	373 248
18	3 24	5 832	73	53 29	389 017
19	3 61	6 859	74	54 76	405 224
20	4 00	8 000	75	56 25	421 875
21	4 41	9 261	76	57 76	438 976
22	4 84	10 648	77	59 29	456 533
23	5 29	12 167	78	60 84	474 552
24	5 76	13 824	79	62 41	493 039
25	6 25	15 625	80	64 00	512 000
26	6 76	17 576	81	65 81	531 441
27	7 29	19 683	82	67 24	551 368
28	7 84	21 952	83	68 89	571 787
29	8 41	24 389	84	70 56	592 704
30	9 00	27 000	85	72 25	614 125
31	9 61	29 791	86	73 96	636 056
32	10 24	32 768	87	75 69	658 503
33	10 89	35 937	88	77 44	681 472
34	11 56	39 304	89	79 21	704 969
35	12 25	42 875	90	81 00	729 000
36	12 96	46 656	91	82 81	753 571
37	13 69	50 653	92	84 64	778 688
38	14 44	54 872	93	86 49	804 357
39	15 21	59 319	94	88 36	830 584
40	16 00	64 000	95	90 25	857 375
41	16 81	68 921	96	92 16	884 736
42	17 64	74 088	97	94 09	912 673
43	18 49	79 507	98	96 04	941 192
44	19 36	85 184	99	98 01	970 299
45	20 25	91 125	100	1 00 00	1 000 000
46	21 16	97 336	101	1 02 01	1 030 301
47	22 09	103 823	102	1 04 04	1 061 208
48	23 04	110 592	103	1 06 09	1 092 727
49	24 01	117 649	104	1 08 16	1 124 864
50	25 00	125 000	105	1 10 25	1 157 625
51	26 01	132 651	106	1 12 36	1 191 016
52	27 04	140 608	107	1 14 49	1 225 043
53	28 09	148 877	108	1 16 64	1 259 712
54	29 16	157 464	109	1 18 81	1 295 029
55	30 25	166 375	110	1 21 00	1 331 000

TABLE OF SQUARES AND CUBES.

SQUARES AND CUBES—*Continued*.

No.	Squares.	Cubes.	No.	Squares.	Cubes.
111	1 23 21	1 367 631	166	2 75 56	4 574 296
112	1 25 44	1 404 928	167	2 78 89	4 657 463
113	1 27 69	1 442 897	168	2 82 24	4 741 632
114	1 29 96	1 481 544	169	2 85 61	4 826 809
115	1 32 25	1 520 875	170	2 89 00	4 913 000
116	1 34 56	1 560 896	171	2 92 41	5 000 211
117	1 36 89	1 601 613	172	2 95 84	5 088 448
118	1 39 24	1 643 032	173	2 99 29	5 177 717
119	1 41 61	1 685 159	174	3 02 76	5 268 024
120	1 44 00	1 728 000	175	3 06 25	5 359 375
121	1 46 41	1 771 561	176	3 09 76	5 451 776
122	1 48 84	1 815 848	177	3 13 29	5 545 233
123	1 51 29	1 860 867	178	3 16 84	5 639 752
124	1 53 76	1 906 624	179	3 20 41	5 735 339
125	1 56 25	1 953 125	180	3 24 00	5 832 000
126	1 58 76	2 000 376	181	3 27 61	5 929 741
127	1 61 29	2 048 383	182	3 31 24	6 028 568
128	1 63 84	2 097 152	183	3 34 89	6 128 487
129	1 66 41	2 146 689	184	3 38 56	6 229 504
130	1 69 00	2 197 000	185	3 42 25	6 331 625
131	1 71 61	2 248 091	186	3 45 96	6 434 856
132	1 74 24	2 299 968	187	3 49 69	6 539 203
133	1 76 89	2 352 637	188	3 53 44	6 644 672
134	1 79 56	2 406 104	189	3 57 21	6 751 269
135	1 82 25	2 460 375	190	3 61 00	6 859 000
136	1 84 96	2 515 456	191	3 64 81	6 967 871
137	1 87 69	2 571 353	192	3 68 64	7 077 888
138	1 90 44	2 628 072	193	3 72 49	7 189 057
139	1 93 21	2 685 619	194	3 76 36	7 301 384
140	1 96 00	2 744 000	195	3 80 25	7 414 875
141	1 98 81	2 803 221	196	3 84 16	7 529 536
142	2 01 64	2 863 288	197	3 88 09	7 645 373
143	2 04 49	2 924 207	198	3 92 04	7 762 392
144	2 07 36	2 985 984	199	3 96 01	7 880 599
145	2 10 25	3 048 625	200	4 00 00	8 000 000
146	2 13 16	3 112 136	201	4 04 01	8 120 601
147	2 16 09	3 176 523	202	4 08 04	8 242 408
148	2 19 04	3 241 792	203	4 12 09	8 365 427
149	2 22 01	3 307 949	204	4 16 16	8 489 664
150	2 25 00	3 375 000	205	4 20 35	8 615 125
151	2 28 01	3 442 951	206	4 24 36	8 741 816
152	2 31 04	3 511 808	207	4 28 49	8 869 743
153	2 34 09	3 581 577	208	4 32 64	8 998 912
154	2 37 16	3 652 264	209	4 36 81	9 129 329
155	2 40 25	3 723 875	210	4 41 00	9 261 000
156	2 43 36	3 796 416	211	4 45 21	9 393 931
157	2 46 49	3 869 893	212	4 49 44	9 528 128
158	2 49 64	3 944 312	213	4 53 69	9 663 597
159	2 52 81	4 019 679	214	4 57 96	9 800 344
160	2 56 00	4 096 000	215	4 62 25	9 938 375
161	2 59 21	4 173 281	216	4 66 56	10 077 646
162	2 62 44	4 251 528	217	4 70 89	10 218 313
163	2 65 69	4 330 747	218	4 75 24	10 360 232
164	2 68 96	4 410 944	219	4 79 61	10 503 459
165	2 72 25	4 492 125	220	4 84 00	10 648 000

TABLE OF SQUARES AND CUBES. 113

SQUARES AND CUBES—*Continued.*

No.	Squares.	Cubes.	No.	Squares.	Cubes.
221	4 88 41	10 793 861	276	7 61 76	21 024 576
222	4 92 84	10 941 048	277	7 67 29	21 253 933
223	4 97 29	11 089 567	278	7 72 84	21 484 952
224	5 01 76	11 239 424	279	7 78 41	21 717 639
225	5 06 25	11 390 625	280	7 84 00	21 952 000
226	5 10 76	11 543 176	281	7 89 61	22 188 041
227	5 15 29	11 697 083	282	7 95 24	22 425 768
228	5 19 84	11 852 352	283	8 00 89	22 665 187
229	5 24 41	12 008 989	284	8 06 56	22 906 304
230	5 29 00	12 167 000	285	8 12 25	23 149 125
231	5 33 61	12 326 391	286	8 17 96	23 393 656
232	5 38 24	12 487 168	287	8 23 69	23 639 903
233	5 42 89	12 649 337	288	8 29 44	23 887 872
234	5 47 56	12 812 904	289	8 35 21	24 137 569
235	5 52 25	12 977 875	290	8 41 00	24 389 000
236	5 56 96	13 144 256	291	8 46 81	24 642 171
237	5 61 69	13 312 053	292	8 52 64	24 897 088
238	5 66 44	13 481 272	293	8 58 49	25 153 757
239	5 71 21	13 651 919	294	8 64 36	25 412 184
240	5 76 00	13 824 000	295	8 70 25	25 672 375
241	5 80 81	13 997 521	296	8 76 16	25 934 336
242	5 85 64	14 172 488	297	8 82 09	26 198 073
243	5 90 49	14 348 907	298	8 88 04	26 463 592
244	5 95 36	14 526 784	299	8 94 01	26 730 899
245	6 00 25	14 706 125	300	9 00 00	27 000 000
246	6 05 16	14 886 936	301	9 06 01	27 270 901
247	6 10 09	15 069 223	302	9 12 04	27 543 608
248	6 15 04	15 252 992	303	9 18 09	27 818 127
249	6 20 01	15 438 249	304	9 24 16	28 094 464
250	6 25 00	15 625 000	305	9 30 25	28 372 625
251	6 30 01	15 813 251	306	9 36 36	28 652 616
252	6 35 04	16 003 008	307	9 42 49	28 934 443
253	6 40 09	16 194 277	308	9 48 64	29 218 112
254	6 45 16	16 387 064	309	9 54 81	29 503 629
255	6 50 25	16 581 375	310	9 61 00	29 791 000
256	6 55 36	16 777 216	311	9 67 21	30 080 231
257	6 60 49	16 974 593	312	9 73 44	30 371 328
258	6 65 64	17 173 512	313	9 79 69	30 664 297
259	6 70 81	17 373 979	314	9 85 96	30 959 144
260	6 76 00	17 576 000	315	9 92 25	31 255 875
261	6 81 21	17 779 581	316	9 98 56	31 554 496
262	6 86 44	17 984 728	317	10 04 89	31 855 013
263	6 91 69	18 191 447	318	10 11 24	32 157 432
264	6 96 96	18 399 744	319	10 17 61	32 461 759
265	7 02 25	18 609 625	320	10 24 00	32 768 000
266	7 06 56	18 821 096	321	10 30 41	33 076 161
267	7 12 89	19 034 163	322	10 36 84	33 386 248
268	7 18 24	19 248 832	323	10 43 29	33 698 267
269	7 23 61	19 465 109	324	10 49 76	34 012 224
270	7 29 00	19 683 000	325	10 56 25	34 328 125
271	7 34 41	19 902 511	326	10 62 76	34 645 976
272	7 39 84	20 123 648	327	10 69 29	34 965 783
273	7 45 29	20 346 417	328	10 75 84	35 287 552
274	7 50 76	20 570 824	329	10 82 41	35 611 289
275	7 56 25	20 796 875	330	10 89 00	35 937 000

SQUARES AND CUBES—*Continued.*

No.	Squares.	Cubes.	No.	Squares.	Cubes.
331	10 95 61	36 264 691	386	14 89 96	57 512 456
332	11 02 24	36 594 368	387	14 97 69	57 960 603
333	11 08 89	36 926 037	388	15 05 44	58 411 072
334	11 15 56	37 250 704	389	15 13 21	58 863 869
335	11 22 25	37 595 375	390	15 21 00	59 319 000
336	11 28 96	37 933 056	391	15 28 81	59 776 471
337	11 35 69	38 272 753	392	15 36 64	60 236 288
338	11 42 44	38 614 472	393	15 44 49	60 698 457
339	11 49 21	38 958 219	394	15 52 36	61 162 984
340	11 56 00	39 304 000	395	15 60 25	61 629 875
341	11 62 81	39 651 821	396	15 68 16	62 099 136
342	11 69 64	40 001 688	397	15 76 09	62 570 773
343	11 76 49	40 353 607	398	15 84 04	63 044 792
344	11 83 36	40 707 584	399	15 92 01	63 521 199
345	11 90 25	41 063 625	400	16 00 00	64 000 000
346	11 97 16	41 421 736	401	16 08 01	64 481 201
347	12 04 09	41 781 923	402	16 16 04	64 964 808
348	12 11 04	42 144 192	403	16 24 09	65 450 827
349	12 18 01	42 508 549	404	16 32 16	65 939 264
350	12 25 00	42 875 000	405	16 40 25	66 430 125
351	12 32 01	43 243 551	406	16 48 36	66 923 416
352	12 39 04	43 614 208	407	16 56 49	67 419 143
353	12 46 09	43 986 977	408	16 64 64	67 917 312
354	12 53 16	44 361 864	409	16 72 81	68 417 929
355	12 60 25	44 738 875	410	16 81 00	68 921 000
356	12 67 36	45 118 016	411	16 89 21	69 426 531
357	12 74 49	45 499 293	412	16 97 44	69 934 528
358	12 81 64	45 882 712	413	17 05 69	70 444 997
359	12 88 81	46 268 279	414	17 13 96	70 957 944
360	12 96 00	46 656 000	415	17 22 25	71 473 375
361.	13 03 21	47 045 881	416	17 30 56	71 991 296
362	13 10 44	47 437 928	417	17 38 89	72 511 713
363	13 17 69	47 832 147	418	17 47 24	73 034 632
364	13 24 96	48 228 544	419	17 55 61	73 560 059
365	13 32 25	48 627 125	420	17 64 00	74 088 000
366	13 39 56	49 027 896	421	17 72 41	74 618 461
367	13 46 89	49 430 863	422	17 80 84	75 151 448
368	13 54 24	49 836 032	423	17 89 29	75 686 967
369	13 61 61	50 243 409	424	17 97 76	76 225 024
370	13 69 00	50 653 000	425	18 06 25	76 765 625
371	13 76 41	51 064 811	426	18 14 76	77 308 776
372	13 83 84	51 478 848	427	18 23 29	77 854 483
373	13 91 29	51 895 117	428	18 31 84	78 402 752
374	13 98 76	52 313 624	429	18 40 41	78 953 589
375	14 06 25	52 734 375	430	18 49 00	79 507 000
376	14 13 76	53 157 376	431	18 57 61	80 062 991
377	14 21 29	53 582 633	432	18 66 24	80 621 568
378	14 28 84	54 010 152	433	18 74 89	81 182 737
379	14 36 41	54 439 939	434	18 83 56	81 746 504
380	14 44 00	54 872 000	435	18 92 25	82 312 875
381	14 51 61	55 306 341	436	19 00 96	82 881 856
382	14 59 24	55 742 968	437	19 09 69	83 453 453
383	14 66 89	56 181 887	438	19 18 44	84 027 672
384	14 74 56	56 623 104	439	19 27 21	84 604 519
385	14 82 25	56 066 625	440	19 36 00	85 184 000

TABLE OF CIRCLES.

THE CIRCUMFERENCE AND AREAS OF CIRCLES FROM 1 TO 50.

Diam.	Circumf.	Area.	Diam.	Circumf.	Area.	Diam.	Circumf.	Area.
1-64	.049087	.00019	**2.**			**5.**		
1-64	.049087	.00019	1-16	6.47953	3.3410	3-16	16.2970	21.135
1-32	.098175	.00077	1-8	6.67588	3.5466	1-4	16.4934	21.648
3-64	.147262	.00173	3-16	6.87223	3.7583	5-16	16.6897	22.166
1-16	.196350	.00307	1-4	7.06858	3.9761	3-8	16.8861	22.691
3-32	.294524	.00690	5-16	7.26493	4.2000	7-16	17.0824	23.221
1-8	.392699	.01227	3-8	7.46128	4.4301	1-2	17.2788	23.758
5-32	.490874	.01917	7-16	7.65763	4.6664	9-16	17.4751	24.301
3-16	.589049	.02761	1-2	7.85398	4.9087	5-8	17.6715	24.850
7-32	.687223	.03758	9-16	8.05033	5.1572	11-16	17.8678	25.406
1-4	.785398	.04909	5-8	8.24668	5.4119	3-4	18.0642	25.967
9-32	.883573	.06213	11-16	8.44303	5.6727	13-16	18.2605	26.535
5-16	.981748	.07670	3-4	8.63938	5.9396	7-8	18.4569	27.109
11-32	1.07992	.09281	13-16	8.83573	6.2126	15-16	18.6532	27.688
3-8	1.17810	.11045	7-8	9.03208	6.4918	**6.**	18.8496	28.274
13-32	1.27627	.12962	15-16	9.22843	6.7771	1-8	19.2423	29.465
7-16	1.37445	.15033	**3.**	9.42478	7.0686	1-4	19.6350	30.680
15-32	1.47262	.17257	1-16	9.62113	7.3662	3-8	20.0277	31.919
1-2	1.57080	.19635	1-8	9.81748	7.6699	1-2	20.4204	33.183
17-32	1.66897	.22166	3-16	10.0138	7.9798	5-8	20.8131	34.472
9-16	1.76715	.24850	1-4	10.2102	8.2958	3-4	21.2058	35.785
19-32	1.86532	.27688	5-16	10.4065	8.6179	7-8	21.5984	37.122
5-8	1.96350	.30680	3-8	10.6029	8.9462	**7.**	21.9911	38.485
21-32	2.06167	.33824	7-16	10.7992	9.2806	1-8	22.3838	39.871
11-16	2.15984	.37122	1-2	10.9956	9.6211	1-4	22.7765	41.282
23-32	2.25802	.40574	9-16	11.1919	9.9678	3-8	23.1692	42.718
3-4	2.35619	.44179	5-8	11.3883	10.321	1-2	23.5619	44.179
25-32	2.45437	.47937	11-16	11.5846	10.680	5-8	23.9546	45.664
13-16	2.55254	.51849	3-4	11.7810	11.045	3-4	24.3473	47.173
27-32	2.65072	.55914	13-16	11.9773	11.416	7-8	24.7400	48.707
7-8	2.74889	.60132	7-8	12.1737	11.793	**8.**	25.1327	50.265
29-32	2.84707	.64504	15-16	12.3700	12.177	1-8	25.5254	51.849
15-16	2.94524	.69029	**4.**	12.5664	12.566	1-4	25.9181	53.456
31-32	3.04342	.73708	1-16	12.7627	12.962	3-8	26.3108	55.088
1.	3.14159	.78540	1-8	12.9591	13.364	1-2	26.7035	56.745
1-16	3.33794	.88664	3-16	13.1554	13.772	5-8	27.0962	58.426
1-8	3.53429	.99402	1-4	13.3518	14.186	3-4	27.4889	60.132
3-16	3.73064	1.1075	5-16	13.5481	14.607	7-8	27.8816	61.862
1-4	3.92699	1.2272	3-8	13.7445	15.033	**9.**	28.2743	63.617
5-16	4.12334	1.3530	7-16	13.9408	15.466	1-8	28.6670	65.397
3-8	4.31969	1.4849	1-2	14.1372	15.904	1-4	29.0597	67.201
7-16	4.51604	1.6230	9-16	14.3335	16.349	3-8	29.4524	69.029
1-2	4.71239	1.7671	5-8	14.5299	16.800	1-2	29.8451	70.882
9-16	4.90874	1.9175	11-16	14.7262	17.257	5-8	30.2378	72.760
5-8	5.10509	2.0739	3-4	14.9226	17.721	3-4	30.6305	74.662
11-16	5.30144	2.2365	13-16	15.1189	18.190	7-8	31.0232	76.589
3-4	5.49779	2.4053	7-8	15.3153	18.665	**10.**	31.4159	78.540
13-16	5.69414	2.5802	15-16	15.5116	19.147	1-8	31.8086	80.516
7-8	5.89049	2.7612	**5.**	15.7080	19.635	1-4	32.2013	82.516
15-16	6.08684	2.9483	1-16	15.9043	20.129	3-8	32.5940	84.541
2.	6.28319	3.1416	1-8	16.1007	20.629	1-2	32.9867	86.590

TABLE OF CIRCLES.
CIRCUMFERENCE AND AREAS OF CIRCLES—*Continued*.

Diam.	Circumf.	Area.	Diam.	Circumf.	Area.	Diam.	Circumf.	Area.
10. 5-8	33.3794	88.664	17. 1-4	54.1925	233.71	23. 7-8	75.0055	447.69
3-4	33.7721	90.763	3-8	54.5852	237.10	24.	75.3982	452.39
7-8	34.1648	92.886	1-2	54.9779	240.53	1-8	75.7909	457.11
11.	34.5575	95.033	5-8	55.3706	243.98	1-4	76.1836	461.86
1-8	34.9502	97.205	3-4	55.7633	247.45	3-8	76.5763	466.64
1-4	35.3429	99.402	7-8	56.1560	250.95	1-2	76.9690	471.44
3-8	35.7356	101.62	18.	56.5487	254.47	5-8	77.3617	476.26
1-2	36.1283	103.87	1-8	56.9414	258.02	3-4	77.7544	481.11
5-8	36.5210	106.14	1-4	57.3341	261.59	7-8	78.1471	485.98
3-4	36.9137	108.43	3-8	57.7268	265.18	25.	78.5398	490.87
7-8	37.3064	110.75	1-2	58.1195	268.80	1-8	78.9325	495.79
12.	37.6991	113.10	5-8	58.5122	272.45	1-4	79.3252	500.74
1-8	38.0918	115.47	3-4	58.9049	276.12	3-8	79.7179	505.71
1-4	38.4845	117.86	7-8	59.2976	279.81	1-2	80.1106	510.71
3-8	38.8772	120.28	19.	59.6903	283.53	5-8	80.5033	515.72
1-2	39.2699	122.72	1-8	60.0830	287.27	3-4	80.8960	520.77
5-8	39.6626	125.19	1-4	60.4757	291.04	7-8	81.2887	525.84
3-4	40.0553	127.68	3-8	60.8684	294.83	26.	81.6814	530.93
7-8	40.4480	130.19	1-2	61.2611	298.65	1-8	82.0741	536.05
13.	40.8407	132.73	5-8	61.6538	302.49	1-4	82.4668	541.19
1-8	41.2334	135.30	3-4	62.0465	306.35	3-8	82.8595	546.35
1-4	41.6261	137.89	7-8	62.4392	310.24	1-2	83.2522	551.55
3-8	42.0188	140.50	20.	62.8319	314.16	5-8	83.6449	556.76
1-2	42.4115	143.14	1-8	63.2246	318.10	3-4	84.0376	562.00
5-8	42.8042	145.80	1-4	63.6173	322.06	7-8	84.4303	567.27
3-4	43.1969	148.49	3-8	64.0100	326.05	27.	84.8230	572.56
7-8	43.5896	151.20	1-2	64.4026	330.06	1-8	85.2157	577.87
14.	43.9823	153.94	5-8	64.7953	334.10	1-4	85.6084	583.21
1-8	44.3750	156.70	3-4	65.1880	338.16	3-8	86.0011	588.57
1-4	44.7677	159.48	7-8	65.5807	342.25	1-2	86.3938	593.96
3-8	45.1604	162.30	21.	65.9734	346.36	5-8	86.7865	599.37
1-2	45.5531	165.13	1-8	66.3661	350.50	3-4	87.1792	604.81
5-8	45.9458	167.99	1-4	66.7588	354.66	7-8	87.5719	610.27
3-4	46.3385	170.87	3-8	67.1515	358.84	28.	87.9646	615.75
7-8	46.7312	173.78	1-2	67.5442	363.05	1-8	88.3573	621.26
15.	47.1239	176.71	5-8	67.9369	367.28	1-4	88.7500	626.80
1-8	47.5166	179.67	3-4	68.3296	371.54	3-8	89.1427	632.36
1-4	47.9093	182.65	7-8	68.7223	375.83	1-2	89.5354	637.94
3-8	48.3020	185.66	22.	69.1150	380.13	5-8	89.9281	643.55
1-2	48.6947	188.69	1-8	69.5077	384.46	3-4	90.3208	649.18
5-8	49.0874	191.75	1-4	69.9004	388.82	7-8	90.7135	654.84
3-4	49.4801	194.83	3-8	70.2931	393.20	29.	91.1062	660.52
7-8	49.8728	197.93	1-2	70.6858	397.61	1-8	91.4989	666.23
16.	50.2655	201.06	5-8	71.0785	402.04	1-4	91.8916	671.96
1-8	50.6582	204.22	3-4	71.4712	406.49	3-8	92.2843	677.71
1-4	51.0509	207.39	7-8	71.8639	410.97	1-2	92.6770	683.49
3-8	51.4436	210.60	23.	72.2566	415.48	5-8	93.0697	689.30
1-2	51.8363	213.82	1-8	72.6493	420.00	3-4	93.4624	695.13
5-8	52.2290	217.08	1-4	73.0420	424.56	7-8	93.8551	700.98
3-4	52.6217	220.35	3-8	73.4347	429.13	30.	94.2478	706.86
7-8	53.0144	223.65	1-2	73.8274	433.74	1-8	94.6405	712.76
17.	53.4071	226.98	5-8	74.2201	438.36	1-4	95.0332	718.69
1-8	53.7998	230.33	3-4	74.6128	443.01	3-8	95.4259	724.64

TABLE OF CIRCLES.

CIRCUMFERENCE AND AREAS OF CIRCLES—Continued.

Diam.	Circumf.	Area.	Diam.	Circumf.	Area.	Diam.	Circumf.	Area.
30. 1-2	95.8186	730.62	37.	116.239	1075.2	43. 1-2	136.659	1486.2
5-8	96.2113	736.62	1-8	116.632	1082.5	5-8	137.052	1494.7
3-4	96.6040	742.64	1-4	117.024	1089.8	3-4	137.445	1503.3
7-8	96.9967	748.69	3-8	117.417	1097.1	7-8	137.837	1511.9
31.	97.3894	754.77	1-2	117.810	1104.5	44.	138.230	1520.5
1-8	97.7821	760.87	5-8	118.202	1111.8	1-8	138.623	1529.2
1-4	98.1748	766.99	3-4	118.596	1119.2	1-4	139.015	1537.9
3-8	98.5675	773.14	7-8	118.988	1126.7	3-8	139.408	1546.6
1-2	98.9602	779.31	38.	119.381	1134.1	1-2	139.801	1555.3
5-8	99.3529	785.51	1-8	119.773	1141.6	5-8	140.194	1564.0
3-4	99.7456	791.73	1-4	120.166	1149.1	3-4	140.586	1572.8
7-8	100.138	797.98	3-8	120.559	1156.6	7-8	140.979	1581.6
32.	100.531	804.25	1-2	120.951	1164.2	45.	141.372	1590.4
1-8	100.924	810.54	5-8	121.344	1171.7	1-8	141.764	1599.3
1-4	101.316	816.86	3-4	121.737	1179.3	1-4	142.157	1608.2
3-8	101.709	823.21	7-8	122.129	1186.9	3-8	142.550	1617.0
1-2	102.102	829.58	39.	122.522	1194.6	1-2	142.942	1626.0
5-8	102.494	835.97	1-8	122.915	1202.3	5-8	143.335	1634.9
3-4	102.887	842.39	1-4	123.308	1210.0	3-4	143.728	1643.9
7-8	103.280	848.83	3-8	123.700	1217.7	7-8	144.121	1652.9
33.	103.673	855.30	1-2	124.093	1225.4	46.	144.513	1661.9
1-8	104.065	861.79	5-8	124.486	1233.2	1-8	144.906	1670.9
1-4	104.458	868.31	3-4	124.878	1241.0	1-4	145.299	1680.0
3-8	104.851	874.85	7-8	125.271	1248.8	3-8	145.691	1689.1
1-2	105.243	881.41	40.	125.664	1256.6	1-2	146.084	1698.2
5-8	105.636	888.00	1-8	126.056	1264.5	5-8	146.477	1707.4
3-4	106.029	894.62	1-4	126.449	1272.4	3-4	146.869	1716.5
7-8	106.421	901.26	3-8	126.842	1280.3	7-8	147.262	1725.7
34.	106.814	907.92	1-2	127.235	1288.2	47.	147.655	1734.9
1-8	107.207	914.61	5-8	127.627	1296.2	1-8	148.048	1744.2
1-4	107.600	921.32	3-4	128.020	1304.2	1-4	148.440	1753.5
3-8	107.992	928.06	7-8	128.413	1312.2	3-8	148.833	1762.7
1-2	108.385	934.82	41.	128.805	1320.3	1-2	149.226	1772.1
5-8	108.778	941.61	1-8	129.198	1328.3	5-8	149.618	1781.4
3-4	109.170	948.42	1-4	129.591	1336.4	3-4	150.011	1790.8
7-8	109.563	955.25	3-8	129.993	1344.5	7-8	150.404	1800.1
35.	109.956	962.11	1-2	130.376	1352.7	48.	150.796	1809.6
1-8	110.348	969.00	5-8	130.769	1360.8	1-8	151.189	1819.0
1-4	110.741	975.91	3-4	131.161	1369.0	1-4	151.582	1828.5
3-8	111.134	982.84	7-8	131.554	1377.2	3-8	151.975	1837.9
1-2	111.527	989.80	42.	131.947	1385.4	1-2	152.367	1847.5
5-8	111.919	996.78	1-8	132.340	1393.7	5-8	152.760	1857.0
3-4	112.312	1003.8	1-4	132.732	1402.0	3-4	153.153	1866.5
7-8	112.705	1010.8	3-8	133.125	1410.3	7-8	153.545	1876.1
36.	113.097	1017.9	1-2	133.518	1418.6	49.	153.938	1885.7
1-8	113.490	1025.0	5-8	133.910	1427.0	1-8	154.331	1895.4
1-4	113.883	1032.1	3-4	134.303	1435.4	1-4	154.723	1905.0
3-8	114.275	1039.2	7-8	134.696	1443.8	3-8	155.116	1914.7
1-2	114.668	1046.3	43.	135.088	1452.2	1-2	155.509	1924.4
5-8	115.061	1053.5	1-8	135.481	1460.7	5-8	155.902	1934.2
3-4	115.454	1060.7	1-4	135.874	1469.1	3-4	156.294	1943.9
7-8	115.846	1068.0	3-8	136.267	1477.6	7-8	156.687	1953.7

THE GOETZ
BOX ANCHOR
AND
POST CAP.

Certainly the most admirable system now known for Slow-Burning Construction.

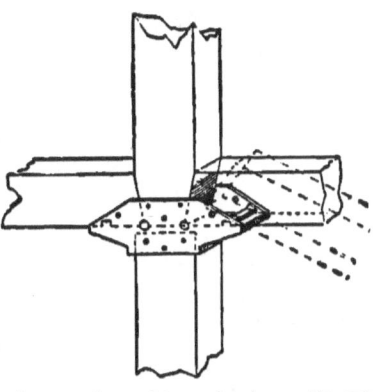

* Posts are fastened to cap forming continuous post from cellar to roof. Falling parts are self-releasing.

In the New Schedule for rating risks, adopted by the National Board of Underwriters, November, 1892, a reduction in rate is made on all buildings that make use of the Goetz methods.

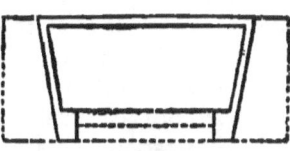

TOP VIEW.

A dovetail box of cast-iron built into the wall; a notch in the girder fits over lug in box, forming a self-releasing anchor.

Showing dovetail form of sides by which the box is locked in the wall.

SEND TO **HOME OFFICE,**
No. 70 STATE STREET, NEW ALBANY, IND.,
for details and catalogue in which this self-releasing method of anchoring is applied to various conditions.

Manufacturing Agents in Every Large City.

SKELETON CONSTRUCTION

AS APPLIED IN

BUILDINGS.

By WM. H. BIRKMIRE.

Fully Illustrated with Engravings from Practical Examples of High Buildings.

8vo, Cloth,

This work includes the description and practical working details of Cast Iron, Wrought Iron, and Steel Columns in the construction of the skeleton frame, and their connections with the Floor and Curtain Wall Girders; Stability of the Structure; Wind Bracing, i.e., Knee and Lateral Bracing; Construction of Joints; Experiments on the Strength of Cast Iron, Wrought Iron, and Steel Columns, such as Z Bar Columns, Phœnix Columns, Plate and Angle Columns, and various commercial rolled shape columns; Floor Framing in the Skeleton Construction.

New York Building Law of 1892 in relation to the Skeleton Frame and Curtain Walls. The same law in relation to the strength of Cast Iron, Wrought Iron, and Steel Columns.

Illustration and Calculation of the Columns, Floor Plans, Tables of Material, Specification, Stairways, Elevators, and Roofs in buildings using Cast Iron, Wrought Iron, and Steel Columns as a skeleton frame, such as "The New Netherlands," a seventeen-story building with nineteen tiers of beams; the Home Life Insurance Building, and others.

To be published in February or March.

FOR SALE BY

JOHN WILEY & SONS,

53 East Tenth Street, New York.

Architectural Iron and Steel,

AND ITS APPLICATION

IN THE

CONSTRUCTION OF BUILDINGS.

INCLUDING BEAMS AND GIRDERS IN FLOOR CONSTRUCTION, ROLLED IRON STRUTS, WROUGHT AND CAST-IRON COLUMNS, FIRE-PROOF COLUMNS, COLUMN CONNECTIONS, CAST-IRON LINTELS, ROOF TRUSSES, STAIRWAYS, ELEVATOR ENCLOSURES, ORNAMENTAL IRON, FLOOR LIGHTS AND SKYLIGHTS, VAULT LIGHTS, DOORS AND SHUTTERS, WINDOW GUARDS AND GRILLES, ETC., ETC., WITH

SPECIFICATION OF IRONWORK.

AND SELECTED PAPERS IN RELATION TO IRONWORK, FROM A REVISION OF THE PRESENT LAW BEFORE THE LEGISLATURE AFFECTING PUBLIC INTERESTS IN THE CITY OF NEW YORK, IN SO FAR AS THE SAME REGULATES THE CONSTRUCTION OF BUILDINGS IN SAID CITY.

TABLES,
SELECTED EXPRESSLY FOR THIS WORK,

OF THE PROPERTIES OF BEAMS, CHANNELS, TEES AND ANGLES, USED AS BEAMS, STRUTS AND COLUMNS, WEIGHTS OF IRON AND STEEL BARS, CAPACITY OF TANKS, AREAS OF CIRCLES, WEIGHTS OF CIRCULAR AND SQUARE CAST-IRON COLUMNS, WEIGHTS OF SUBSTANCES, TABLES OF SQUARES, CUBES, ETC., WEIGHTS OF SHEET COPPER, BRASS AND IRON, ETC.

BY

WM. H. BIRKMIRE,
OF J. B. & J. M. CORNELL IRON WORKS, 141 CENTRE STREET.

Fully Illustrated.

8VO, CLOTH, ... $3.50.

NEW YORK:
JOHN WILEY & SONS,
53 East Tenth Street.
1891.

CONTENTS.

CHAPTER I.

THE MANUFACTURE OF IRON.

ARTICLE	PAGE
1. Iron,	1
2. Smelting,	1
3. Pig Iron,	1

WROUGHT IRON.

4. Puddling,	2
5. Piling,	2
6. Rolling,	3
7. Channels,	3
8. Quality of Wrought Iron,	3
9. Testing Wrought Iron,	4
10. Cold Bend Test,	4
11. Modulus of Elasticity,	5
12. Wrought Iron in Compression,	5
13. Weight of Wrought Iron,	5

STEEL.

14. Steel,	5
15. Mild or Soft Steel,	6
16. Rolling Steel,	6
17. Billets,	6
18. Weight of Steel,	7

CAST IRON.

19. Cast Iron,	7
20. Castings,	8
21. Cores,	8
22. Crushing Strength of Cast Iron,	8
23. Tenacity of Cast Iron,	8
24. Weight of Cast Iron,	8

CHAPTER II.

FLOORS.

ARTICLE	PAGE
25. Dead Load,	9
26. Live Load,	9
27. Method of determining Rolled-iron Beams by Diagram,	9
28. Beams as Girders,	11
29. To determine Coefficient for Beams,	11
30. Properties of Wrought-iron I Beams,	12
31. Deflection,	12
32. Coefficients for Steel Beams,	13
33. Properties of Steel I Beams,	13
34. Channels,	14
35. Properties of Wrought-iron Channels,	14
36. Properties of Steel Channels,	15
37. Zee Bars,	15
38. Floors should be Rigid,	15
39. Elastic Limit,	15
40. Maximum Deflection,	16
41. Framed Beams,	16
42. Tie Rods,	16
43. Beam Connections,	18
44. Bearing for Beams,	18
45. Pressure on Brick and Stone Work,	18
46. Knees for Beam Connections,	18
47. Bolts and Rivets for Beam Connections,	19
48. Tee Irons as Beams,	19
49. Angle Irons as Beams (Even Legs),	20
50. Angle Irons as Beams (Uneven Legs),	20
51. Beams not uniformly loaded, and Beams not supported at both Ends,	20

CHAPTER III.

GIRDERS.

52. Compound Girders,	22
53. Webs,	22
54. Buckling,	22
55. Flanges,	22
56. Deflection,	22
57. Rivets in Girders,	22
58. Strain on Flanges of Girders,	23
59. Shearing,	24
60. Flanges reduced in Thickness near Ends,	24
61. Weight of Brickwork,	26
62. Separators,	27

CONTENTS.

ARTICLE	PAGE
63. Cast-iron Plates on Girders for Walls,	28
64. Bolts and Rivets,	28
65. Shearing and Bearing of Rivets,	28
66. Pins,	29

CHAPTER IV.

CAST-IRON LINTELS.

67. Cast-iron Lintels,	30
68. Skew-back Lintels,	31
69. Lintels for Iron Fronts,	31
70. Sidewalk Lintels,	32
71. Window-head Lintels,	32
72. Double-web Lintels,	32
73. Window Sills,	32
74. Rule for Breaking Weight at Middle,	32
75. Webs,	33
76. Tests of Cast-iron Lintels,	33

CHAPTER V.

TRUSSES.

77. Roof Trusses,	35
78. Loads on Trusses,	35
79. Snow and Wind Pressure,	36
80. Ceiling Weight,	36
81. The Graphic Method,	36
82. King-post Truss,	36
83. Truss No. 2,	38
84. Truss No. 3,	38
85. Truss No. 4,	39
86. Truss No. 5,	40
87. Details of Iron Trusses,	42
88. Ties and Struts,	42
89. Wooden Purlins,	42
90. Connections,	42

CHAPTER VI.

STRUTS.

91. Rolled-iron Struts,	44
92. End Connections,	44
93. Factors of Safety,	46

viii   CONTENTS.

ARTICLE		PAGE
94.	Greatest Safe Load on Struts,	47
95.	Channel Struts,	48
96.	Angles as Struts,	48
97.	Tees as Struts,	49
98.	Properties of Beams for Struts,	50
99.	Properties of Channels for Struts,	51
100.	Properties of Angles for Struts,	52
101.	Properties of Tees for Struts,	54

CHAPTER VII.

CAST-IRON COLUMNS.

102.	Columns, Shafts Ornamented,	55
103.	Capitals,	55
104.	Cast-iron Column Connection,	56
105.	Holes Drilled,	58
106.	Column Flanges,	58
107.	Fire-proof Column,	58
108.	Dowel Columns,	59
109.	Strength of Cast-iron Columns,	60
110.	Weight to be Estimated for any Use,	62
111.	Strength of Hollow Cast-iron Columns,	62
112.	Factors of Safety for Cast-iron Columns,	63
113.	Ribbed Base Plates,	64
114.	Flat Base Plates,	65
115.	Grouting,	65
116.	Bedding,	66
117.	Cast-iron Dowels for Wooden Columns,	66
118.	Wrought-iron Pins and Cast-iron Star-shaped Dowels,	66

CHAPTER VIII.

WROUGHT-IRON COLUMNS.

119.	Wrought-iron Column Sections,	67
120.	Zee-bar Columns,	68
121.	Strength of Wrought-iron Columns,	68
122.	Safe Load on Wrought-iron Columns,	69
123.	Radii of Gyration for Round Column,	70
124.	Radii of Gyration for Square Column,	70

CHAPTER IX.

STAIRWAYS.

125.	Close-string Stairs,	71
126.	Height and Breadth of Steps,	71
127.	Cast-iron Stairs,	71

CONTENTS. ix

ARTICLE		PAGE
128.	To Measure Height of Railing,	72
129.	Number of Strings regulated by Width,	74
130.	Wrought-iron Stairs,	74
131.	Circular Stairs,	76
132.	Deck-beam Strings,	77
133.	Channel Strings,	77
134.	Plate and Angle Strings,	77
135.	Treads and Risers,	78
136.	Fascias,	79
137.	Posts or Newels,	79
138.	Brackets for Stair Handrail,	79
139.	Stairs to be carefully constructed,	79

CHAPTER X.

ORNAMENTAL IRONWORK.

140.	Ornamental Design,	80
141.	Hammered Wrought Iron,	80
142.	Method of Hammering Leaves, etc.,	83
143.	Hammered Wrought-iron Grilles,	83
144.	Cast-iron Ornamented,	83
145.	Modelling for Ornamental Castings,	84
146.	Finish of Ornamental Iron,	86

CHAPTER XI.

ELEVATOR ENCLOSURES.

147.	Passenger-elevator Enclosure,	88
148.	Freight-elevator Enclosure,	90
149.	Double Sliding Doors for Passenger-elevator Fronts,	92
150.	Elevator Guide Supports,	94

CHAPTER XII.

DOORS AND SHUTTERS.

151.	Circular-head Door and Frame,	95
152.	Sidewalk Door,	95
153.	Outside Folding Shutters,	98
154.	Shutter Hinges,	99
155.	Storm Hooks,	99
156.	Shutter Eyes,	99
157.	Shutter Rings,	99
158.	Cast-iron Brick,	99
159.	Rolling Steel Shutters,	100
160.	The Noiseless Shutter,	100

CHAPTER XIII.
FLOOR LIGHTS AND SKYLIGHTS.

ARTICLE	PAGE
161. Cast-iron Floor Lights for Iron Beams,	102
162. Cast-iron Floor Lights for Wooden Beams,	102
163. Wrought-iron Floor Lights for Wooden Beams,	102
164. Skylights,	106
165. Hip Skylight,	109

CHAPTER XIV.
HOLLOW BURNT CLAY.

166. Hollow Blocks for Arches,	110
167. Porous Terra Cotta,	110

CHAPTER XV.
ANCHORS.

168. Ashler Anchors,	113
169. Side Anchors,	113
170. Wall Anchors,	114
171. Hook Anchors,	114
172. Drive Anchors,	114
173. Wedge Anchors,	115
174. Coping Anchors,	116
175. Government or V Anchors,	116
176. Girder Straps,	116
177. Beam Straps,	116

CHAPTER XVI.
BOLTS.

178. Square-head Bolts,	117
179. Hexagon-head Bolts,	117
180. Button-head or Carriage Bolts,	117
181. Countersunk-head Bolts,	117
182. Screw-head Bolts,	117
183. Tap Bolts,	117
184. Counter-sunk Tap Bolts,	118
185. Double and Single Expansion Bolts,	118
186. Lag Screws,	119
187. Upset Ends,	119
188. Open-drop Forged Turn-buckles, Pipe Swivel and Arm Swivel,	119

CHAPTER XVII.
MISCELLANEOUS DETAILS.

189. Mail Chutes,	120
190. Folding Gates,	122

CONTENTS.

ARTICLE		PAGE
191.	Box Slides,	123
192.	Hanging Ceilings,	124
193.	Flitch-plate Girders,	125
194.	Sidewalk Elevator,	127
195.	Wrought-iron Gratings,	129
196.	Cast-iron Perforated Plates,	130
197.	Knee Gratings,	130
198.	Galvanized Iron Cornices,	130
199.	Scuttle,	131
200.	Scuttle Ladder,	131
201.	Iron Fronts,	132
202.	Plain Fire Escapes,	134
203.	Brackets on New Buildings,	135
204.	Top Rails,	136
205.	Bottom Rails,	136
206.	Filling-in Bars,	136
207.	Stairs,	136
208.	Floors,	136
209.	Drop Ladders,	136
210.	Height of Railing,	137
211.	Ornamental Fire Escapes,	137
212.	Fire Escapes for Schools, Factories, etc.,	137
213.	Vault Cover and Frame,	137
214.	Cast-iron Grating,	138
215.	Strainers,	138
216.	Plain Wrought-iron Bar Window Guards,	138
217.	Bridle or Stirrup Irons,	139
218.	Chimney Cap,	142
219.	Cast-iron Flue Door and Frame,	142
220.	Wrought-iron Flue Door,	142
221.	Flue Ring and Cover,	142
222.	Chimney Ladder,	143
223.	Corrugated Iron,	143
224.	Galvanized and Black Iron,	144
225.	Finial and Crestings,	145
226.	Vault Lights,	146
227.	Wrought-iron Guard,	147
228.	Dwarf Doors,	148
229.	Cast-iron Wheel Guards,	148
230.	Fire Pipes,	148
231.	Mansard Roof,	148
232.	Railings for Roof Protection,	148
233.	Pipe Railing,	149
234.	Corrugated Flooring,	149
235.	Tanks,	150
236.	Chains and Cables,	152

xii CONTENTS.

ARTICLE	PAGE
237. Wire Work,	154
238. Cast-iron Boiler Flues,	155
239. Wrought-iron Boiler Flues,	155

CHAPTER XVIII.

FINISHING IRON AND STEEL.

240. Bronzing,	156
241. Enamelling Cast and Wrought iron,	156
242. Electro-plating,	156
243. Galvanizing Sheet Iron,	157
244. Painting of Iron,	157
245. Malleable Castings,	158
246. Lacquer for Iron,	158

CHAPTER XIX.

SPECIFICATION.

General Conditions,	159
Time of Completion,	160
Payments,	160
Constructive Work,	161
Wrought Iron,	161
Wrought Steel,	161
Cast Iron,	162
Painting,	162
Anchors, Clamps, Dowels, etc.,	162
Columns,	162
Girders,	163
Cast-iron Lintels,	164
Stairs,	164
Bulkheads on roof,	165
Miscellaneous,	165
Setting,	165

CHAPTER XX.

TABLES.

Average Weight in Pounds of a Cubic Foot of Various Substances, 166, 167 168
Squares and Cubes, of Numbers from 1 to 440, . . . 169, 170, 171, 172
The Circumference and Areas of Circles from 1 to 50, . . . 173, 174, 175
Weight per Foot of Flat Iron, ... 176, 177
Number of U. S. Gallons (231 Cubic Inches) Contained in Circular Tanks, . 178

CONTENTS.

	PAGE
Decimal Equivalents for Fractions of an Inch,	178
Decimal Equivalents for Fractions of a Foot,	179
Weight of 100 Bolts with Square Heads and Nuts,	180
Weights of Nuts and Bolt-heads in Pounds,	180
Weight of Sheets of Wrought Iron, Steel, Copper, and Brass,	181
Weight of Square Cast-iron Columns in Pounds per Lineal Foot,	182
Weight per Lineal Foot of Circular Cast-iron Columns,	183

CHAPTER XXI.

NEW YORK BUILDING LAW.

Vault Lights and Areas Protected,	184
Buildings Increased in Size by Use of Columns and Girders,	185
Anchors,	185
Floors, Stairs and Ceilings of Iron,	185
Weight on Floors,	186
Framing of Beams,	187
Cast-iron Templates,	188
Iron Lintels,	188
Fire-proof Columns,	189
Iron Fronts Backed with Brick,	190
Thickness of Cast-iron Posts,	190
Curtain-wall Girders,	190
Rolled Iron and Steel Beams and Factors of Safety,	191
Girders to be Tested,	192
Beams, Lintels and Girders to be Inspected,	192
Stirrup Irons,	193
Beam Anchors,	193
Flitch-plates for Girders,	193
Smoke Flues Lined with Cast Iron,	194
Iron Shutters,	194
Railings around Well-holes,	195
Elevator Wells Inclosed with Brick or Iron,	196
Dumb-waiters, Skylights over Elevators,	196
Screen of Iron under Elevator Machinery,	196
Mansard Roof,	196
Bulkheads,	197
Cornices and Gutters,	197
Dormer Windows, Scuttles and Skylights,	197
Iron Ladders to Scuttles,	198
Fire Escapes,	198
Roof Gardens on Theatres,	198
Proscenium Wall Girder, etc.,	199
Skylights and Doorways to Theatres,	199

	PAGE
Doorways through Proscenium Wall, Doors of Iron and Wood,	199
Roof of Auditorium, Main Floor of Auditorium and Vestibule, Floor of Second Story over Entrance, Lobby and Corridors of Iron and Fire-proof Materials,	200
The Fronts of Galleries, Ceiling of Auditorium, Partitions in Auditorium, Entrance Vestibule, Partitions of Dressing-rooms and Doors in same to be Fire-proof,	200
Actors' Dressing-rooms and Fly Galleries,	201
Stage and Fly Galleries Fire-proof,	201
Proscenium Opening and Curtain,	201

www.ingramcontent.com/pod-product-compliance
Lightning Source LLC
Chambersburg PA
CBHW030320170426
43202CB00009B/1081